Grade 3

Reveal MATH®

Student Practice Book

Mc
Graw
Hill

mheducation.com/prek-12

Send all inquiries to:
McGraw Hill
8787 Orion Place
Columbus, OH 43240

ISBN: 978-0-07-693706-6
MHID: 0-07-693706-2

Printed in the United States of America.

8 9 10 11 LHS 26 25 24 23

Table of Contents

Unit 2

Use Place Value to Fluently Add and Subtract within 1,000

Lessons

Unit 3

Multiplication and Division

Lessons

Unit 4
Use Patterns to Multiply by 0, 1, 2, 5, and 10

Lessons

Unit 5
Use Properties to Multiply by 3, 4, 6, 7, 8, and 9

Lessons

Unit 6
Connect Area and Multiplication

Lessons

Unit 7
Fractions
Lessons

Unit 8
Fraction Equivalence and Comparison
Lessons

Unit 9
Use Multiplication to Divide

Lessons

Unit 10
Use Properties and Strategies to Multiply and Divide

Lessons

Unit 11
Perimeter

Lessons

Unit 12
Measurement and Data

Lessons

Unit 13
Lessons

Describe and Analyze 2-Dimensional Shapes

Additional Practice

Name _____

Review

There are patterns in sums when the addends are even and odd numbers.

When you add two even numbers, the sum is even.

$348 + 204 = 552$ $124 + 236 = 360$ $572 + 420 = 992$

When you add two odd numbers, the sum is even.

$421 + 123 = 544$ $615 + 187 = 802$ $259 + 301 = 560$

When you add an even number and an odd number, the sum is odd.

$602 + 157 = 759$ $517 + 322 = 839$ $243 + 406 = 649$

What makes the statement true? Write *even* or *odd*. Then write 2 equations using 3-digit numbers to support your answer.

1. _____ + odd = even

2. odd = odd + _____

3. even + _____ = even

What is the sum? Use patterns to help justify your answer.

4. 312 + 287 = _____

5. 135 + 453 = _____

_____

6. A piece of David's homework accidentally tore off. As his teacher was grading his work, she could see only that David wrote 43 as the last two digits of the sum 532 + 100. How can the teacher know that David's work is incorrect without looking at the hundreds place?

7. A screen on Evelyn's cell phone can hold an odd or an even number of apps. If she has an odd number of apps, how can she arrange them on 2 screens?

Math @ Home Activity

Roll three dice (or one die three times) and record the die values as a 3-digit number. For example, if 4, 6, and 2 are rolled, record 462. Do this twice. Have your child determine if the sum of the two numbers is even or odd.

Lesson **2-7**

Additional Practice

Name ______________________________

Review

You can decompose one number in a subtraction problem to find the difference.

Decompose using place value.

417 − 266 = ?
417 − 200 = 217
217 − 60 = 157
157 − 6 = 151

Decompose another way.

417 − 266 = ?
417 − 217 = 200
200 − 40 = 160
160 − 9 = 151

How can you decompose the number in 2 ways?

1. 629

2. 583

How can you decompose one number to subtract? Why did you choose that way?

3. 696 − 275

4. 726 − 340

How can you find the difference? Show the strategy you used.

5. 536 − 234 = _____

6. 854 − 426 = _____

7. 904 − 684 = _____

8. 623 − 363 = _____

9. A baker bakes 487 muffins for an order. 273 are banana muffins. The rest are blueberry muffins. How many blueberry muffins does she bake?

10. Ryan subtracts 739 − 574 by decomposing 574. She subtracts 4, then subtracts 500, and then subtracts 70. Will her answer be correct? Explain your reasoning.

Math @ Home Activity

Identify two house or building numbers in your neighborhood. Have your child subtract the two numbers using decomposition (using only the last 3 digits of the numbers if necessary).

Additional Practice

Name _____

Review

You can adjust numbers in addition and subtraction equations to make the equation easier to work with.

Adjust Addition Equations

$$513 + 172 = ?$$

-3   +3

$$510 + 175 = 685$$

Subtract from one addend and add that amount to the other addend.

Adjust Subtraction Equations

$$369 - 125 = ?$$

-5   -5

$$364 - 120 = 244$$

Subtract from or add the same amount to both numbers.

How can you adjust the equation by the given amount and solve it?

1. $362 - 142 = ?$ Adjust by adding 3.

2. $654 + 261 = ?$ Adjust by adding and subtracting 4.

How can you adjust the equation to solve?

3. $524 - 219 = ?$

4. $622 + 207 = ?$

5. $873 - 528 = ?$

6. $432 + 534 = ?$

7. Tianyu and Marissa are finding $477 + 239$. Tianyu finds the sum by rewriting the expression as $480 + 236$. Marissa claims that Tianyu's expression is incorrect. She says the sum should be found by rewriting the expression as $476 + 240$. Is Marissa correct? Explain.

Provide your child with subtraction and addition problems that use page numbers of a book he or she is reading. Encourage your child to explain the strategy used to find the difference or the sum.

Additional Practice

Name _____

Review

You can use bar diagrams to represent situations involving addition and subtraction.

Brooke makes programs for a school play. She needs a total of 675 programs. She has made 340 programs. Use a bar diagram to represent this situation. How many more programs does Brooke need to make?

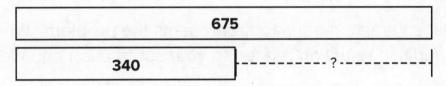

Write a subtraction and addition equation to represent the situation.

$675 - 340 = ?$ $340 + ? = 675$

$675 - 340 = 335$ $340 + 335 = 675$

Complete the problem.

1. Which equations are related to $736 - 314 = 422$? Circle all that apply.

   A. $422 + 736 = 314$ **C.** $314 + 422 = 736$

   B. $736 - 422 = 314$ **D.** $736 + 314 = 422$

2. Which equations are related to $672 - 230 = 442$? Circle all that apply.

   A. $230 + 442 = 672$ **C.** $672 - 442 = 230$

   B. $672 + 230 = 442$ **D.** $442 + 230 = 672$

3. Braxton has 460 trading cards. He gives 323 cards to his brother. Which equation can Braxton use to find how many trading cards he has left?

323 trading cards	----? trading cards----
460 trading cards	

A. $460 + ? = 323$

C. $323 - ? = 460$

B. $460 + 323 = ?$

D. $460 - 323 = ?$

---

4. A pet store has 235 fish for sale. In one day, they sell 140 fish. How many fish are left?

What subtraction equation represents the problem? What is an addition equation related to your subtraction equation?

_____ − _____ = _____

_____ + _____ = _____

5. Mrs. Walker has 480 books in her classroom. She gives 185 books to a new teacher. How many books does Mrs. Walker have left?

What subtraction equation represents the problem? What is an addition equation related to your subtraction equation?

_____ − _____ = _____

_____ + _____ = _____

Give your child two small handfuls of coins. Count the number of coins in each handful with your child. Have him or her write an addition equation to represent the total, followed by a related subtraction equation.

Additional Practice

Name _____

Review

You can use different strategies to find the sum when adding.

Partial Sums Use place value to decompose each addend.

$$527 + 288 = ?$$
$$500 + 200 = 700$$
$$20 + 80 = 100$$
$$7 + 8 = 15$$
$$700 + 100 + 15 = 815$$

$$500 + 200$$
$$20 + 80$$
$$7 + 8$$

$$
\begin{array}{r}
527 \\
+\ 288 \\
\hline
700 \\
100 \\
15 \\
\hline
815
\end{array}
$$

Adjust Addends Adjust addends to make them easier to add. Subtract from one addend and add that amount to the other.

$$527 + 288$$
$$-2 \quad\ +2$$
$$525 + 290 = 815$$

How can you find the sum?

1. $172 + 399 =$ _____

2. $509 + 411 =$ _____

3. $\begin{array}{r} 667 \\ +\ 219 \\ \hline \end{array}$

4. $\begin{array}{r} 574 \\ +\ 406 \\ \hline \end{array}$

How can you find the sum? Explain your strategy choice.

5. $692 + 265 = $ _____

6. $443 + 534 = $ _____

_____

7. Jacob and Raul race to a tree 359 feet away and then to a fence 242 feet away. How far do they race in all? Show how you found your answer.

8. Sarah is adding $171 + 258$. She adds 2 to 258 to add $260 + 171$. Then she adds 2 to the sum. Do you agree with her strategy? Explain.

9. Nikki biked 315 miles in June and 387 miles in July. How far did Nikki bike in June and July combined? Show how you found your answer.

Math @ Home Activity

On three index cards write *Decompose Both Addends Using Place Value, Adjust the Addends* and *Any Strategy*. Give your child two 3-digit numbers to add. Have your child chose one of the strategies to find the sum, then explain why he or she used that strategy.

Additional Practice

Name _____

Review

You can use bar diagrams to solve two-step problems.

Charlie has $810. He pays a $220 bill and a $365 bill. How much money does Charlie have left after he pays these two bills?

Step 1 Determine how much money Charlie needs to pay bills.

You can use an addition equation.

$$220 + 365 = a$$

total of bills
a

| 220 | 365 |

$$220 + 365 = 585$$
Charlie pays $585.

Step 2 Determine how much money Charlie has left.

You can use a subtraction equation.

$$810 - 585 = b$$

money left
| 585 | b |
| 810 |

$$810 - 585 = 225$$
Charlie will have $225 left.

How can you write an equation to represent the bar diagram?

1.

| 531 cars | 487 cars |

a

_____

2.

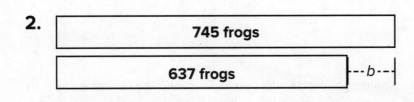

_____

Represent and solve the problem. Use letters for the unknowns.

3. Blakely grows 847 zucchini. She sells 215 zucchini. She gives away 140 zucchini. How many zucchini does she have left?

4. Tisha collects stamps. She has 612 stamps. Her mother gives her 131 more stamps. She then sells 107 of her stamps. How many stamps does she have now?

5. Victor is giving out flyers for a sporting event. He gave out 368 flyers. Then, he was given 248 more flyers to give out. If Victor now has 875 flyers, how many flyers did he start with?

Write a two-step word problem for your child involving a family task such as shopping or paying bills. Have him or her explain the steps needed to find the solution.

Additional Practice

Name _____

Review

You can multiply the number of objects in each group by the number of equal groups to find the total number of objects.

If Jay buys five 4-packs of batteries, he buys a total of 20 batteries. 5 × 4 = 20.

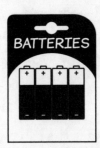 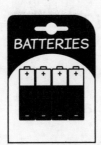   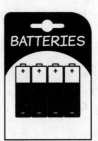

5 packs × 4 batteries per pack = 20 batteries in all

How can you use a drawing to represent the equal groups?

1. 4 equal groups of 6

2. 5 equal groups of 2

3. 2 equal groups of 8

4. What multiplication equation represents the equal groups?

5. Haley buys markers in packages of 4. How many markers are in 3 packages?

   a. How can you draw a picture to represent the problem?

   b. What equation represents the problem?

   c. What is the solution? Fill in the blank.

     There are _____ markers in 3 packages.

6. Randy earns money from walking dogs. He earns $5 for walking each of 6 dogs. How much does Randy earn? Explain how you know.

7. Finn fills an order for boxes of nails at 4 constructions sites. He orders the same number of boxes for each site. How many boxes of nails might he order? Explain how you know.

Find things around your home that come in packages, such as grocery items or batteries. Have your child write multiplication equations to find the total number of objects in a certain number of packages.

Copyright © McGraw-Hill Education

Additional Practice

Name _____

Review

According to the Commutative Property of Multiplication, you can change the order of the factors in a multiplication equation and the product stays the same.

The arrays show that 3 × 5 has the same product as 5 × 3. Both 3 × 5 and 5 × 3 equal 15.

Draw representations to show that the products are equal.

1. 2 × 3 and 3 × 2

2. 8 × 3 and 3 × 8

3. 5 × 4 and 4 × 5

4. 6 × 7 and 7 × 6

5. Avery says that 2 × 7 and 7 × 2 have the same product. Is Avery correct? Explain.

6. Brett has 3 packs of 4 pens each, and Lindsey has 4 packs of 3 pens each. Draw equal groups to show that Brett and Lindsey have the same number of pens. What multiplication equation matches what each person has?

Brett: _____	Lindsey: _____

7. Arturo cut 6 pieces of rope that are each 5 feet long. Kim cut 5 pieces of rope. If they both cut the same total amount of rope, how long were the pieces that Kim cut? Explain.

8. Thomas needs to set up 35 chairs in equal rows for a talent show. He sets up 5 equal rows of 7 chairs. Is there another way he can set up the chairs in equal rows? Explain.

Look for arrays of objects around your home with your child. For example, you might find a dresser with 2 columns of 3 drawers each, ceiling tiles in 4 rows of 5, or a muffin tin with 3 rows of 4 cups. For each array you find, have your child write two multiplication equations.

Additional Practice

Name _____

Review

You can divide by sharing objects equally among groups.

Start with 18 counters and equally share them among
3 groups. Each group gets 6 counters. 18 ÷ 3 = 6.

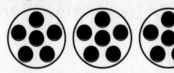

How can you draw a representation for each equation? How can you complete the equation?

1. 30 ÷ 6 = _____   **2.** 14 ÷ 2 = _____

3. 20 ÷ 5 = _____   **4.** 21 ÷ 3 = _____

5. How can you write a division equation for the representation?

_____

Draw a representation. Then solve the problem.

6. There are 16 balloons for a celebration. The balloons are shared equally among 4 tables. How many balloons are at each table?

There are _____ balloons at each table.

7. Jaxon has 12 bracelets to share among his 4 friends. How many bracelets does each friend get?

Each friend gets _____ bracelets.

8. A hotel has 28 rooms that need to be cleaned. If 4 housekeepers each clean the same number of rooms, will there be any rooms left over? Justify your answer.

Use groups of objects around your home to help your child practice division by equal sharing. Starting with a number of objects, ask your child to determine how many each person would receive when the objects are shared equally among a certain number of people. Be sure that the total number of objects is a multiple of the number of sharing groups to avoid having a remainder. Your child can make groups with the objects to help divide, and then write a division equation that describes the groups.

Additional Practice

Name _____

Review

You can divide by separating objects into groups of equal size.

Create equal groups of 8 until you reach a total of 24. The representation shows that there are 3 groups of 8 in 24, so $24 \div 8 = 3$.

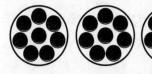

What division equation represents the equal grouping?

1.   2.

3.

4.

Draw a representation. Then write an equation to describe the problem.

5. There are 20 pencils in a box. Each student gets 2 pencils. How many students are there?

There are _____ students.

6. Sherice has 24 books to give away. If Sherice wants to give 3 books to each friend, how many friends will receive books?

Sherice can give books to _____ friends.

7. Emil makes 21 baked goods for a bake sale. He puts an equal number of baked goods on each plate. How many plates does Emil need? Justify your answer.

Math @ Home Activity

Have your child write math riddles about different animals and the number of legs they have. For example, "Some Emperor penguins are huddled in a group for warmth. They have a total of 16 legs. How many penguins are there?" Have your child determine the answer by drawing equal groups. Then have him or her write a division equation to match the equal grouping.

Additional Practice

Name _____

Review

You can use arrays or equal groups to show how multiplication and division are related.

The array can be used to write related multiplication and division equations.

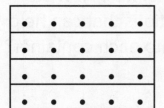

4 groups of 5 = 20   20 divided by 4 = 5

4 × 5 = 20   20 ÷ 5 = 4

How can you draw equal groups for the equations?

1. 3 groups of 3 = 9

9 divided by 3 = 3

2. 4 × 2 = 8

8 ÷ 4 = 2

How can you draw an array for the equations?

3. 5 groups of 2 = 10

10 divided by 5 = 2

4. 2 × 7 = 14

14 ÷ 2 = 7

What multiplication and division equations can you write for the representation?

5.    _____

_____

6. How can you write a related division equation?

$6 \times 3 = 18$

7. How can you write a related mutliplication equation?

$12 \div 4 = 3$

---

8. Henry divides 90 nails equally into 10 containers. How can you write multiplication and division equations that can help Henry determine how many nails to put into each container?

9. Ms. Miller has 56 sheets of color paper to give to her students. There are 8 students, and each student gets the same amount of paper. How can you use the multiplication equation $8 \times 7 = 56$ to find out how many sheets of paper each student should get?

10. Raelyn has 33 red beads and 21 blue beads. She mixes the beads together and uses 6 beads for each key chain she makes. What multiplication equation and division equation can help Raelyn determine how many key chains she can make?

Math @ Home Activity

Write multiplication and corresponding division equations on index cards. Have your child identify matching pairs of index cards that have related multiplication and division equations on them. For example, the equations $3 \times 4 = 12$ and $12 \div 3 = 4$ are a match. To make it more challenging, remove one number from each equation and replace it with the ? symbol.

Additional Practice

Name _____

Review

You can use what you know about multiplying 5s to solve problems.

Nora is decorating a wall with picture frames. She arranges the frames into 5 columns and some rows. How many frames could Nora have in all?

Products of 5 have a 0 or 5 in the ones place. So, Nora could have 5, 10, 15, 20, 25, 30, 35, 40, 45, or 50 frames.

What number completes the equation?

1. $5 \times$ _____ $= 40$

2. $5 \times 3 =$ _____

3. _____ $\times 5 = 25$

4. $45 = 5 \times$ _____

5. $5 \times 7 =$ _____

6. _____ $\times 5 = 30$

7. $5 \times 10 =$ _____

8. _____ $\times 5 = 10$

---

9. Which equations are true? Choose all that apply.

A. $5 \times 4 = 5 + 5 + 5 + 5$ **B.** $5 \times 5 = 5 + 5$

C. $5 \times 5 = 5 + 5 + 5 + 5 + 5$ **D.** $5 \times 6 = 5 + 6$

10. Which equation is true?

   A. $5 \times 1 = 5 + 1$

   B. $5 \times 2 = 5 + 5$

   C. $5 \times 3 = 5 + 3$

   D. $5 \times 4 = 5 + 5 + 5 + 5 + 5$

How can you use what you know about multiplication to solve?

11. How does knowing 5×4 help you remember 4×5?

12. Mark is organizing his DVDs into piles. Each pile has 5 DVDs. How many DVDs could Mark have in all? Explain.

13. Ming has 5 piles of fiction books. Each pile has 3 books. How many fiction books does Ming have?

   Ming has _____ fiction books.

14. An apple stand has 5 bags of apples. There are 6 apples in each bag. How many apples are on the apple stand?

   There are _____ apples on the apple stand.

Math @ Home Activity

Have your child trace his or her hand once and then write a multiplication fact to represent the number of fingers ($5 \times 1 = 5$). Repeat activity with different numbers of hands drawn.

Additional Practice

Name _____

Review

You can use what you know about multiplication with 10 to solve problems.

Each children's book has 10 pages. How many total pages are there in 4 children's books?

$4 \times 10 = 40$. There are 40 pages.

---

1. Isabel has 6 sets of crayons. There are 10 crayons in each set. What pattern can you use to find the total number of crayons?

2. Marvin has 10 baskets of fruit. He fills each basket with the same number of apples. There are fewer than 11 apples in each basket. How many apples could Marvin have? Explain.

3. Which equations are true? Choose all that apply.

   A. $4 \times 5 = 4 \times 10$ **B.** $10 + 10 + 10 = 10 \times 3$

   C. $10 + 10 + 10 = 10 \times 10$ **D.** $5 \times 6 = 3 \times 10$

4. Draw a line to the number that makes each equation true.

_____ = 3 × 10	70
10 × _____ = 60	30
2 × _____ = 20	5
_____ = 7 × 10	10
_____ × 10 = 50	40
4 × 10 = _____	6

How can you use what you know about multiplication with 10 to answer the question?

5. What number is in the ones place of a product of 10 and another number? Explain.

6. How does knowing 3 × 5 help you remember 3 × 10?

7. Tony bikes 10 laps each day. How many laps does Tony bike in 4 days?

Tony bikes _____ laps.

Math @ Home Activity

Have your child use dimes to practice the ×10 facts. Have your child count a small handful of dimes and then write a coordinating multiplication fact. Repeat with different amounts of dimes.

Additional Practice

Name _____

Review

You can use what you know about multiplying with 1 and 0 to solve problems.

Bryce bikes 1 mile a day for 10 days. How many miles does he bike altogether?

$$10 \times 1 = 10$$

Any number multiplied by 1 equals itself.

Simon puts 0 books in 2 boxes. How many books does he place in the boxes?

$$2 \times 0 = 0$$

Any number multiplied by 0 equals 0.

What number completes the equation?

1. $8 \times 1 =$ _____

2. $9 = 1 \times$ _____

3. _____ $= 0 \times 10$

4. $0 \times 3 =$ _____

5. $0 =$ _____ $\times 4$

6. _____ $\times 1 = 7$

7. $5 \times$ _____ $= 5$

8. _____ $\times 6 = 0$

---

9. Zack has a container of strawberries. He opens the container and realizes that there are no strawberries left. How many strawberries does Zack get?

   Zack gets _____ strawberries.

10. Which equations are true? Choose all that apply.

A. $2 \times 0 = 2$ **B.** $5 \times 0 = 9 \times 0$

C. $6 \times 1 = 3 \times 2$ **D.** $1 \times 3 = 3$

11. There are 10 fish bowls at a pet store. There is 1 fish in each fish bowl. How many fish are there?

12. June starts reading a new book. She reads 0 pages each day for 8 days. How many pages will June have read by the end of 8 days?

June will have read _____ pages.

13. Ms. Andrea is ordering tights for each dancer in the eight-year-old class. She orders the same number of tights as dancers.

What equation could you write to describe the number of tights Ms. Andrea could order? Explain.

14. Miguel says that 4×0 and 6×0 have the same product. Is he correct? Explain.

Math @ Home Activity

Provide opportunities for your child to practice multiplying by 1 and 0. When setting the table, ask your child how many cups, plates, etc. are needed for a certain number of people. Include items for which 1 of each is needed and for which 0 of each is needed.

Additional Practice

Name _____

Review

You can identify patterns to help you solve multiplication problems.

The product of 0 and any number is always 0.

$$9 \times 0 = 0, 0 \times 6 = 0$$

Products of 2 have a 0, 2, 4, 6, or 8 in the ones place.

$$2 \times 4 = 8, 5 \times 2 = 10$$

The product of any number and 1 is the other number.

$$1 \times 3 = 3, 5 \times 1 = 5$$

Products of 5 have a 0 or 5 in the ones place.

$$5 \times 8 = 40, 7 \times 5 = 35$$

When you multiply a digit by 10, the product has the same digit in the ten place and 0 in the ones place.

$$7 \times 10 = 70, 10 \times 9 = 90$$

What word or number makes each statement true?

1. Products of 2 are always _____.

2. By using number patterns, you can tell that 50 is the product of _____ and _____ because it has a 0 in the ones place.

3. By using number patterns, you can tell that 45 is a product of _____ because it has a 5 in the ones place.

4. By using number patterns, you can tell that 0 is the product of _____ and _____ because any number multiplied by _____ is 0.

5. Sara has fewer than ten $5 bills in her wallet. How much money might she have?

6. A clothing store employee is making a display of pairs of socks. Can the employee use 17 socks in all? Explain.

7. Mike staples 10 sheets of paper together to make a packet. He makes fewer than 8 packets. He uses all of the paper. How many sheets of paper could Mike have used? Explain how you know.

8. Is it possible for a number to be a product of 2, a product of 5, and a product of 10? Explain.

Math @ Home Activity

Allow time for your child to practice identifying and using patterns when multiplying. For example, while sorting change, have your child use the patterns of 1, 5, and 10 to determine the amount of money present.

Additional Practice

Name _____

Review

You can use decomposition to find the total number of counters.

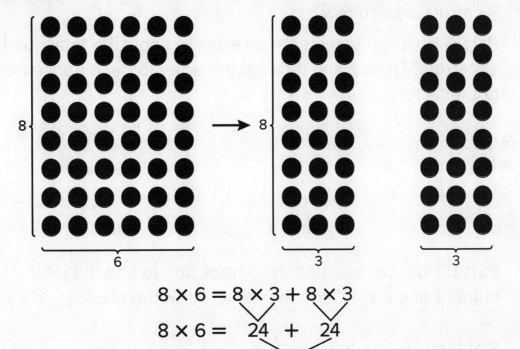

$$8 \times 6 = 8 \times 3 + 8 \times 3$$
$$8 \times 6 = \quad 24 \; + \; 24$$
$$8 \times 6 = \qquad 48$$

How can one factor be decomposed to find the product?

1. $5 \times 8 =$ _____ × _____ + _____ × _____

$5 \times 8 =$ _____ + _____

$5 \times 8 =$ _____

2. $9 \times 3 =$ _____ × _____ + _____ × _____

$9 \times 3 =$ _____ + _____

$9 \times 3 =$ _____

How can one factor be decomposed to find the product?

3. 6 × 7 = _____ × _____ + _____ × _____

   6 × 7 = _____ + _____

   6 × 7 = _____

4. Carmen plants 8 rows of carrots in her garden. She plants 9 carrots in each row.

   Part A How can you create an array to show the total number of carrots? Draw a line through the array to decompose one of the factors.

   Part B How can you use decomposition to find the total number of carrots Carmen plants? Fill in the blanks.

   8 × 9 = _____ × _____ + _____ × _____

   8 × 9 = _____ + _____

   8 × 9 = _____

   Carmen plants _____ carrots.

Math @ Home Activity

Give your child a multiplication equation, such as 7 × 4 = ?, to solve. Have him or her use dry pasta to create an array to represent the multiplication equation. Ask your child to decompose one of the factors in the array to solve the equation. Then have him or her decompose the other factor in the array to solve the equation. Compare the products.

Additional Practice

Name _____

Review

You can decompose 4s facts into two 2s facts. Find the product of one 2s fact. Then double the product.

Chad reads 9 pages a day. How many pages does he read in 4 days?

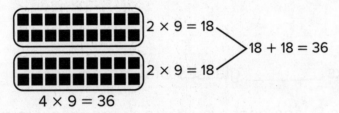

$2 \times 9 = 18$

$2 \times 9 = 18$

$18 + 18 = 36$

$4 \times 9 = 36$

$9 \times 4 = 36$. Chad reads 36 pages in 4 days.

What number completes the equation?

1. $5 \times$ _____ $= 20$

2. $8 \times 4 =$ _____

3. $40 = 4 \times$ _____

4. $3 \times$ _____ $= 12$

5. _____ $\times 6 = 24$

6. $7 \times 4 =$ _____

7. $4 \times 2 =$ _____

8. _____ $\times 4 = 36$

9. $4 \times 1 =$ _____

10. $4 \times 4 =$ _____

11. _____ $\times 7 = 28$

12. $4 \times 5 =$ _____

How can you use 2s fact to find the unknown?

13. $6 \times 2 = 12$ $6 \times 4 = $ _____

14. $2 \times 3 = 6$ $4 \times 3 = $ _____

15. $5 \times 2 = 10$ $5 \times 4 = $ _____

How can you use a 2s fact to solve the problem? Show your thinking.

16. Tori drinks 4 glasses of water each day. How many glasses of water does Tori drink in 7 days?

Tori drinks _____ glasses of water in 7 days.

17. Zach runs 8 laps each day at track practice. How many laps does Zach run in 4 days?

Zach runs _____ laps in 4 days.

18. Li Bo found the product of 4×9. He took the following steps:

• First, he found the product of 2×3, which is 6.
• Then, he added $6 + 6$ to get 12.
He says that $4 \times 9 = 12$. What would you say to LiBo? Explain your reasoning.

Have your child make 4 groups of beads. Help your child use the sum of two 2s facts to write an equation that represents the groups. Repeat with different numbers of beads in each group.

Additional Practice

Name _____

Review

You can solve multiplication problems with 6 using different strategies.

You can decompose the factor 6 into 5 and 1.

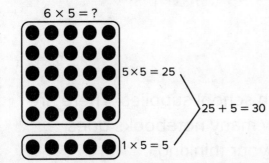

You can decompose the factor 6 into 3 and 3.

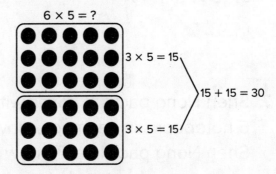

How can you use the 3s facts to find the unknown?

1. $3 \times 7 = 21$ $6 \times 7 =$ _____

2. $3 \times 4 = 12$ _____ $\times 4 = 24$

3. $6 \times 3 = 18$ $6 \times$ _____ $= 36$

4. Coach Green puts 8 volleyballs in each of 6 carts. How many volleyballs are there in all?

There are _____ volleyballs in all.

5. Which expression is equivalent to 4 × 6?

    A. 3 × 1 + 3 × 4   **B.** 3 × 2 + 3 × 2

    C. 5 × 4 + 1 × 4   **D.** 5 × 2 + 5 × 2

6. Which expressions are equivalent to 6 × 9? Choose all that apply.

    A. 3 × 9 + 3 × 9   **B.** 3 × 5 + 3 × 4

    C. 5 × 9 + 1 × 1   **D.** 9 × 5 + 9 × 1

7. Shen Nong packs 6 boxes with school supplies. There are 6 notebooks in each box. How many notebooks does Shen Nong pack in all? Show your thinking.

8. Bruce says that if a number is a product of 6, then it is *not* a product of 3. What would you say to Bruce?

Look for situations in which your child can recall multiplication facts with 6. Have your child use a strategy to find the product. Then have him or her explain the relationship between the strategy and the 6s fact.

Additional Practice

Name _____

Review

You can decompose to find 7s and 9s facts. You can use 5s facts to help recall a 7s or 9s fact.

Decompose 7 into 5 and 2 to find the product of 7×6.

Decompose 9 into 5 and 4 to find the product of 9×6.

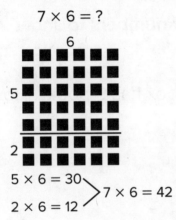

$7 \times 6 = ?$

$5 \times 6 = 30$
$2 \times 6 = 12$
$7 \times 6 = 42$

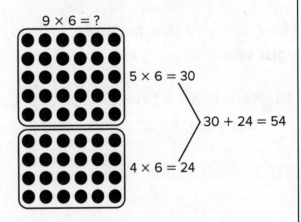

$9 \times 6 = ?$

$5 \times 6 = 30$
$4 \times 6 = 24$
$30 + 24 = 54$

What number completes the equation?

1. $9 \times 5 =$ _____

2. $56 = 7 \times$ _____

3. _____ $\times 3 = 27$

4. $49 =$ _____ $\times 7$

5. $9 \times$ _____ $= 81$

6. $7 \times 6 =$ _____

7. _____ $\times 9 = 72$

8. $5 \times 7 =$ _____

9. $63 = 9 \times$ _____

10. Which expression is equivalent to 9×6?

A. $1 \times 9 + 6 \times 9$

B. $2 \times 6 + 5 \times 6$

C. $4 \times 6 + 5 \times 6$

D. $5 \times 3 + 4 \times 3$

11. Which expression is equivalent to 7×8?

A. $2 \times 8 + 5 \times 8$

B. $3 \times 3 + 5 \times 8$

C. $5 \times 4 + 5 \times 4$

D. $5 \times 4 + 2 \times 4$

How can you decompose using 5s facts to determine the product of 7 × 9? Complete each step.

12. 7 × 9 = ?

5 × _____ = 45

2 × 9 = _____

45 + _____ = _____

7 × 9 = _____

13. 7 × 9 = ?

_____ × 7 = 35

4 × 7 = _____

35 + _____ = _____

7 × 9 = _____

How can you use pictures, words, and numbers to show your work?

14. Kerri buys 8 boxes of books. Each box has 9 books. How many books does she buy in all?

15. Changan has 7 recordings on his phone. Each one is 6 minutes long. How many minutes of recordings are on his phone?

Math @ Home Activity

Look for situations in which your child can practice multiplying by 7s or 9s using 5s facts. For example, give your child a 7s fact. Have him or her use 5s facts to find the product: 7 × 3 = 5 × 3 + 2 × 3 = 15 + 6 = 21

Additional Practice

Name _____

Review

You can find the area of a figure by covering it with unit squares with no gaps or overlaps.

Is this figure tiled correctly to find the area?

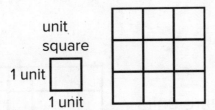

Yes, it is covered with square units with no gaps or overlaps. It has an area of 9 square units.

Is this figure tiled correctly to find the area?

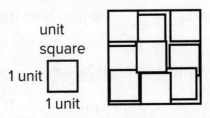

No, it is covered with square units but there are gaps and overlaps. You cannot use this model to find the area.

---

1. Which figure is tiled correctly to find the area?

A.

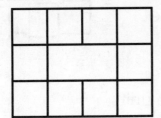

B.

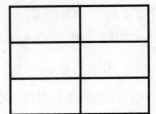

C.

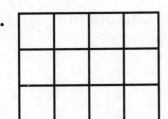

D.

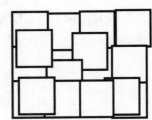

How can you complete the tiling and find the area of the figure?

2.

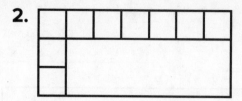

$A =$ _____ square units

3.

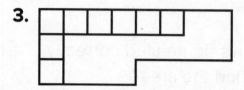

$A =$ _____ square units

4. What is the area of this figure?

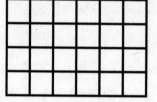

   A. 4 square units   **B.** 6 square units

   C. 20 square units   **D.** 24 square units

5. Manon tiles a figure like this. She says, "It has
14 square units."

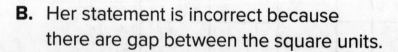

How would you respond to Manon?
Choose all that apply.

   A. Her statement is incorrect because
the squares units are too small.

   B. Her statement is incorrect because
there are gap between the square units.

   C. Her statement is incorrect because the square units overlap.

   D. Her statement is incorrect because she did not
use square units.

Math @ Home Activity

Cut small squares from index cards. Have your child arrange some or all
of the squares to make different figures. Then have your child count the
squares to find the area of the figure.

Additional Practice

Name _____

Review

You can multiply the length of a rectangle by its width to find the area.

Use a multiplication equation to represent the area.

4 rows of unit squares

✕ 6 unit squares in each row
_____

24 unit squares

The area is 24 square units.

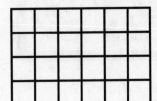

What is the area of each rectangle?

1. _____ square _____

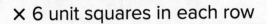

7 mm

4 mm   4 mm

7 mm

2. _____ square _____

9 in.

4 in.   4 in.

9 in.

3. _____ square _____

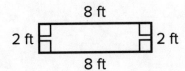

8 ft

2 ft   2 ft

8 ft

4. _____ square _____

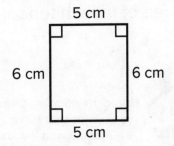

5 cm

6 cm   6 cm

5 cm

What is the area of each object?

5.

⟵ 3 ft ⟶

4 ft

The area of the
painting is _____
square _____.

6.

9 in.

6 in.

The area of the cutting
board surface is _____
square _____.

7. Demetrius gets a new
whiteboard to use for
homework. The dimensions
of the whiteboard are shown.
What is the area of the
whiteboard?

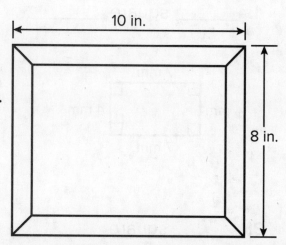

10 in.

8 in.

The area of the whiteboard is _____.

Write the numbers 2 through 10 on index cards. Shuffle the cards and have
your child randomly choose two cards. Then have your child find the area of
a rectangle with length and width equal to the numbers on the cards.

Additional Practice

Name _____

Review

You can find the area of a rectangle by decomposing it into two smaller rectangles and adding their areas together.

You can use a multiplication equation to find the area of the rectangle.

$19 \times 6 = ?$

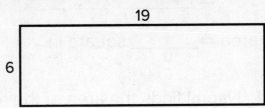

You can make two smaller rectangles to make the equation easier to solve.

$19 \times 6 = 10 \times 6 + 9 \times 6$

$\qquad = 60 + 54$

$\qquad = 114$

The area is 114 square units.

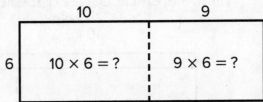

How can you decompose to find the area of each rectangle?

1.

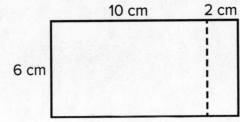

2.

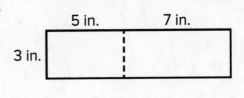

$12 \times 6 =$ ____ $\times 6 +$ ____ $\times 6$

$\qquad =$ ____ $+$ ____

area = ____ square cm

$12 \times 3 =$ ____ $\times 3 +$ ____ $\times 3$

$\qquad =$ ____ $+$ ____

area = ____ square in.

3. How can you decompose the rectangle into two smaller rectangles to find the area of the original rectangle?

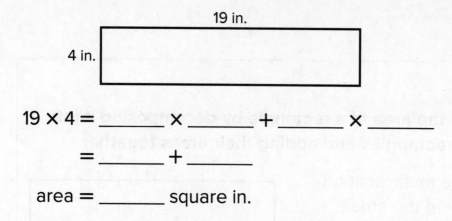

$19 \times 4 = $ _____ $\times$ _____ $+$ _____ $\times$ _____

$= $ _____ $+$ _____

area $= $ _____ square in.

4. Naomi finds the area of the rectangle. Her work is shown.

$4 \times 18 = 4 \times 10 + 4 \times 6 + 4 \times 2$

Will the area be correct? Explain.

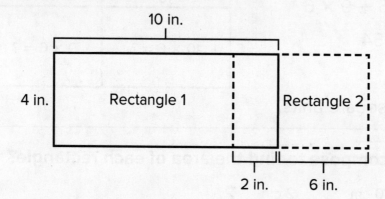

Additional Practice

Name _____

Review

You can use representations to make sense of area problems. One strategy to solve the problem is to decompose the figure.

Ada plants a flower garden. What is the area of her garden?

You can decompose the figure into 2 rectangles. Multiply the length and width of each rectangle.

Add the areas.

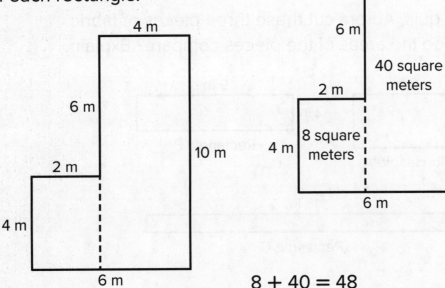

$4 \times 2 = 8$ $10 \times 4 = 40$

$8 + 40 = 48$

The area is 48 square meters.

---

1. Elsa buys a rug for her bedroom. It is 12 feet long and 6 feet wide. What is the area of the rug?

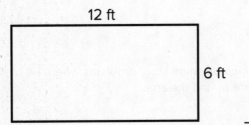

_____ square feet

2. Kylo designs a space in a park for picnics and playgrounds. What is the area of the space he designs?

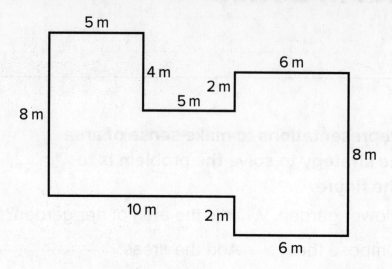

_____ square meters

3. For a quilt, Aurora cut these three pieces of fabric. How do the areas of the pieces compare? Explain.

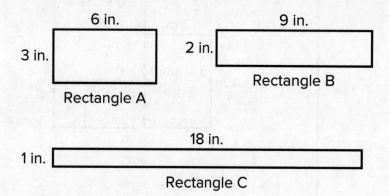

Additional Practice

Name _____

Review

You can represent fractions on number lines using fraction intervals.

Use the denominator to determine how to partition the number line into equal parts. Each interval represents $\frac{1}{8}$.

Use the numerator to find how many intervals to count. Count by eighths 5 times. The point represents $\frac{5}{8}$.

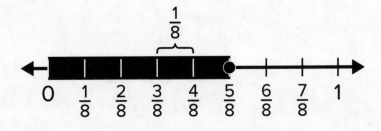

How can you fill in the fraction labeled with a point?

1.

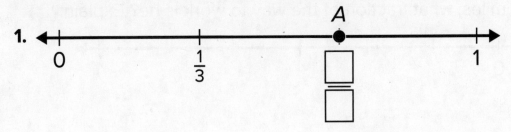

2.

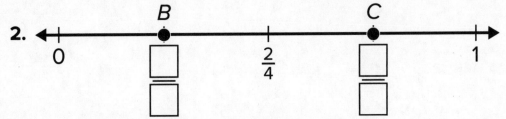

3.

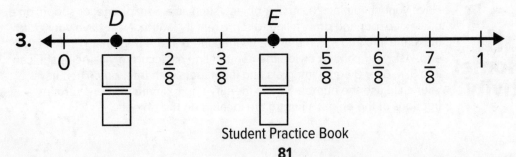

4. What fraction is represented by point *A*?

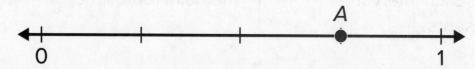

---

5. Alleah is painting all four sides of a shed. Each side of the shed is the same size. She placed Point *A* on the number line to represent the fraction of the shed that she has painted so far. What fraction of the shed has she painted? Explain.

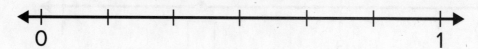

6. Rhett drives 6 miles to work each day. When he has driven 5 miles, what fraction of the way to work is he? Explain.

Have your child place a smaller object, such as a pen, phone, or spoon, on a piece of paper and draw a line to represent its length. Next have your child make the line into a number line by marking 0 at the beginning and 1 at the end. Ask your child to divide the number line as accurately as he or she can into 3, 4, 6, or 8 equal intervals and label them with fractions. Then have your child use the number line to identify approximately where certain fractions of the object's length are located on the object.

Additional Practice

Name _____

Review

When the numerator and denominator of a fraction are the same, you can determine the fraction is equal to one whole, or 1.

The numerator and denominator of $\frac{3}{3}$ are the same, so $\frac{3}{3}$ is equal to 1.

What fraction represents the shaded part of the shape?

1.

☐
─
☐

2.

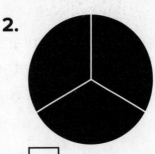

☐
─
☐

3.

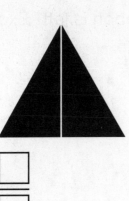

☐
─
☐

4.

☐
─
☐

5. How can you label the number line using fractions? What fraction represents 1?

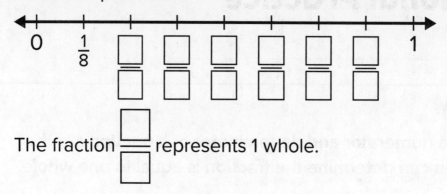

The fraction $\dfrac{\square}{\square}$ represents 1 whole.

6. How can you model $\dfrac{4}{4} = 1$ in two different ways?

7. Rochelle cuts a pan of lasagna into 6 equal pieces. She puts 1 piece on each of 6 plates. Which fraction represents the amount of lasagna that Rochelle used? Explain.

8. Alan cuts a long ribbon into 8 equal pieces. He uses $\dfrac{8}{8}$ of the ribbon to make bows. How much of the ribbon is left? Explain.

Try coming up with a rhyme, song, acronym, or other mnemonic device with your child to help him or her remember when a fraction is equal to 1. Try to include the words *numerator* and *denominator*, if possible.

Additional Practice

Name _____

Review

You can represent any whole number as a fraction.

Marco has 4 fruit strips. How can you represent the number of fruit strips he has as a fraction?

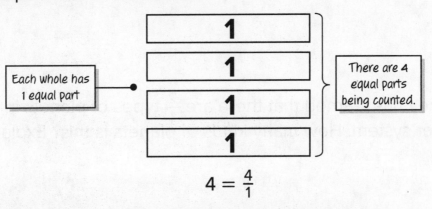

Each whole has 1 equal part

There are 4 equal parts being counted.

$$4 = \frac{4}{1}$$

What fraction represents the whole number?

1.   $\dfrac{\square}{\square} = 2$

2.   $\dfrac{\square}{\square} = 3$

3. $\dfrac{\square}{\square} = 4$

4. Which fractions are equal to a whole number? Choose all that apply.

    A. $\frac{1}{8}$ **B.** $\frac{6}{1}$ **C.** $\frac{3}{4}$

    D. $\frac{5}{1}$ **E.** $\frac{2}{3}$ **F.** $\frac{1}{4}$

5. Breanna has 8 jugs of lemonade to sell. How can you express the number of jugs of lemonade as a fraction? Explain.

6. Haley has learned that there are $\frac{2}{1}$ types of planets in our solar system. How many kinds of planets is this? Explain.

7. If each moon is a whole, why is the number of moons not $\frac{1}{8}$?

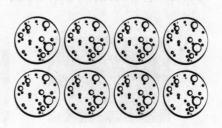

Additional Practice

Name _____

Review

You can represent equivalent fractions.

Fractions that represent the same amount of same-sized wholes are equivalent.

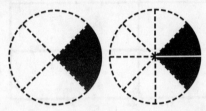

$\frac{1}{4}$ and $\frac{2}{8}$ are equivalent fractions. Both fractions shade the same amount of same-sized wholes.

How can you shade the model to show the equivalent fractions?

1.

2.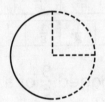

3.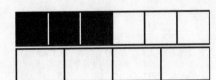

4.

5. How can you use the fraction models to determine if the fractions are equivalent? Choose equivalent or not equivalent.

	Equivalent	Not Equivalent
$\frac{2}{3}$ and $\frac{2}{6}$		
$\frac{4}{8}$ and $\frac{1}{2}$		
$\frac{3}{4}$ and $\frac{6}{8}$		
$\frac{2}{2}$ and $\frac{3}{3}$		
$\frac{5}{6}$ and $\frac{5}{8}$		

$\frac{1}{2}$		$\frac{1}{2}$		

$\frac{1}{3}$	$\frac{1}{3}$	$\frac{1}{3}$

$\frac{1}{4}$	$\frac{1}{4}$	$\frac{1}{4}$	$\frac{1}{4}$

$\frac{1}{6}$	$\frac{1}{6}$	$\frac{1}{6}$	$\frac{1}{6}$	$\frac{1}{6}$	$\frac{1}{6}$

$\frac{1}{8}$	$\frac{1}{8}$	$\frac{1}{8}$	$\frac{1}{8}$	$\frac{1}{8}$	$\frac{1}{8}$	$\frac{1}{8}$	$\frac{1}{8}$

6. Jared jogged $\frac{3}{6}$ of a mile. Elliot jogged $\frac{7}{8}$ of a mile. Janelle jogged $\frac{2}{3}$ of a mile. Karis jogged $\frac{4}{8}$ of a mile. Which two students jogged the same distance? Explain your answer.

Math @ Home Activity

Help your child identify equivalent fractions. Fold paper into two equal pieces and shade one of the parts. Have your child write the fraction represented by the shaded parts. Then fold paper into 4, 6, or 8 equal parts. Have your child write the new equivalent fractions. Try it again. Fold the paper into equal parts. Shade a few. Then fold the paper again to create an equivalent fraction.

Additional Practice

Name _____

Review

You can determine whether fractions are equivalent by seeing if they represent the same point on number lines when the wholes are the same length.

The fractions $\frac{1}{2}$ and $\frac{2}{4}$ represent the same point on a number line, so $\frac{1}{2}$ is equivalent to $\frac{2}{4}$.

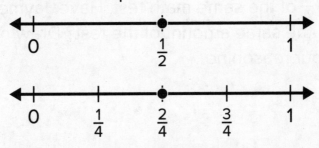

---

1. How can you use the points on the number lines to name the equivalent fractions?

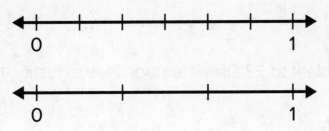

$\dfrac{\square}{6} = \dfrac{\square}{\square}$ $\dfrac{\square}{6} = \dfrac{\square}{\square}$

$\dfrac{\square}{6} = \dfrac{\square}{\square}$ $\dfrac{\square}{6} = \dfrac{\square}{\square}$

2. How can you use the number lines to determine whether $\frac{3}{6}$ is equivalent to $\frac{4}{8}$? Explain your work.

3. Jaymee has completed $\frac{2}{8}$ of a math test and Darius has completed $\frac{1}{4}$ of the same math test. Have Jaymee and Darius completed the same amount of the test? Draw number lines to justify your reasoning.

4. Is $\frac{5}{6}$ equivalent to $\frac{6}{8}$? Draw number lines to justify your reasoning.

Write fractions with denominators 2, 3, 4, 6, and 8 on index cards. For example, write $\frac{1}{6}$, $\frac{2}{6}$, $\frac{3}{6}$, $\frac{4}{6}$, and $\frac{5}{6}$ for the denominator 6. Have your child choose two cards at random and use number lines to determine whether the fractions are equivalent.

Additional Practice

Name _____

<table>
<tr><td colspan="2">

Review

You can compare fractions if the wholes are the same size and shape.
</td></tr>
<tr>
<td>

You cannot compare $\frac{1}{3}$ of a rectangle to $\frac{1}{3}$ of a circle because the wholes are not the same shape.

</td>
<td>

You cannot compare $\frac{1}{3}$ of a small rectangle to $\frac{1}{3}$ of a large rectangle because the wholes are not the same size.

</td>
</tr>
</table>

1. Are the parts equivalent? Choose Yes or No.

	Yes	**No**

2. Do the fraction models represent the same amount? Explain.

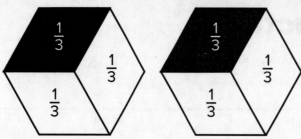

3. How can you draw two models of $\frac{3}{4}$ that do not represent the same amount?

4. Boris puts water for his cat in a small, round bowl. He puts water for his dog in a large, round bowl. After the animals drink, each bowl is $\frac{1}{8}$ full of water. Do they contain the same amount? Explain.

Math @ Home Activity

Provide your child with different sizes and shapes of paper. Have him or her fold the papers into eighths. Shade the same number of parts on each paper and compare the amount shaded.

Additional Practice

Name _____

Review

You can compare fractions with the same denominator by comparing their numerators.

Since 3 parts of a whole is greater than 2 parts of the same whole, the fraction $\frac{3}{4}$ is greater than $\frac{2}{4}$.

$\frac{2}{4}$
| $\frac{1}{4}$ | $\frac{1}{4}$ | $\frac{1}{4}$ | $\frac{1}{4}$ |

$\frac{3}{4}$
| $\frac{1}{4}$ | $\frac{1}{4}$ | $\frac{1}{4}$ | $\frac{1}{4}$ |

When comparing fractions with the same denominator, the fraction with the greater numerator is greater.

Which symbol >, <, or = will make the comparison true? Shade the fraction tiles to justify your reasoning.

1. $\frac{5}{6} \bigcirc \frac{3}{6}$

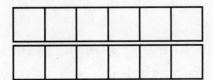

2. $\frac{4}{8} \bigcirc \frac{1}{8}$

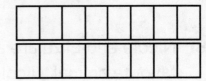

3. $\frac{2}{4} \bigcirc \frac{3}{4}$

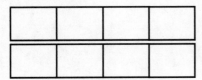

4. $\frac{1}{3} \bigcirc \frac{2}{3}$

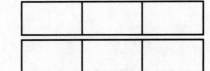

Which symbol >, <, or = will make the comparison true? Draw a point on the number lines to justify your reasoning.

5. $\frac{3}{8}$ ◯ $\frac{2}{8}$

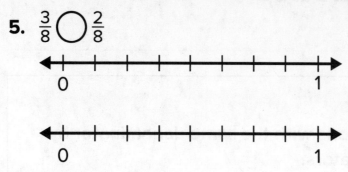

6. $\frac{1}{6}$ ◯ $\frac{5}{6}$

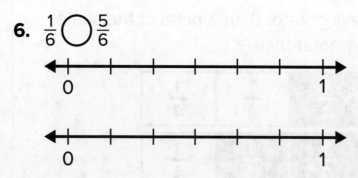

7. Rick and Leah use the same size of paper to write a story. Rick writes a story that is $\frac{5}{8}$ page long. Leah writes a story that is $\frac{4}{8}$ page long. Who writes more? Use pictures or words to explain your reasoning.

8. Nicole and Eliza are asking their classmates to name their favorite animal. Nicole asked $\frac{2}{4}$ of the class. Eliza asked fewer people than Nicole. What fraction of the class could Eliza have asked? Explain your reasoning.

Math @ Home Activity

Cut index cards into equal pieces: one card into equal pieces of 3, one card into equal pieces of 4, one card into equal pieces of 6, and one card into equal pieces of 8. Label the pieces accordingly with $\frac{1}{3}$, $\frac{1}{4}$, $\frac{1}{6}$, and $\frac{1}{8}$. Next, write two fractions with the same denominator (3, 4, 6, or 8). Have your child compare the fractions using >, <, or = and then test the comparison by representing each fraction with the index card pieces.

Additional Practice

Name _____

Review

You can compare fractions using number lines and fraction models.

The denominators are the same. So, the fraction with the greater numerator is greater.

$$\frac{1}{3} < \frac{2}{3}$$

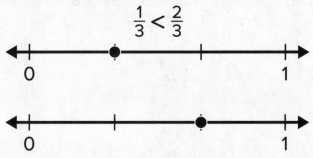

The numerators are the same. So, the fraction with the lesser denominator is greater.

$$\frac{2}{3} > \frac{2}{4}$$

$\frac{1}{3}$	$\frac{1}{3}$	$\frac{1}{3}$

$\frac{1}{4}$	$\frac{1}{4}$	$\frac{1}{4}$	$\frac{1}{4}$

Which can you use >, <, or = to make the comparison true? Draw number lines or fraction models to justify your answer.

1. $\frac{6}{6} \bigcirc \frac{2}{2}$

2. $\frac{3}{4} \bigcirc \frac{3}{8}$

3. $\frac{1}{2}$ ◯ $\frac{2}{2}$

4. $\frac{2}{6}$ ◯ $\frac{2}{3}$

5. Which comparisons are true? Circle them. Explain your reasoning.

$\frac{2}{4} > \frac{2}{3}$ $\frac{1}{6} < \frac{5}{6}$ $\frac{2}{3} > \frac{2}{8}$ $\frac{3}{8} > \frac{7}{8}$

6. Olivia's work and Opal's work are shown. How can you draw a representation of each girl's work to determine if either made a mistake?

Olivia	Opal
$\frac{3}{3} > \frac{3}{4}$	$\frac{5}{8} > \frac{5}{6}$

Math @ Home Activity

Provide opportunities for your child to compare fractions. Give him or her two fractions to compare. Have him or her draw representations to compare the fractions and justify the comparison.

Additional Practice

Name _____

Review

You can use the relationship between multiplication and division to represent a division equation as an unknown factor problem.

Fact triangles can help you rewrite a division equation as an unknown factor problem. $$32 \div 4 = ?$$ $$? \times 4 = 32$$ 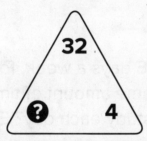	The quotient and unknown factor are the same number. Finding the unknown factor will give you the quotient. $$8 \times 4 = 32$$ $$32 \div 4 = 8$$ 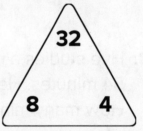

How can you use a fact triangle to find a related multiplication equation and the unknown number? Complete the equation and the fact triangle.

1. $21 \div 3 =$ _____

   $3 \times$ _____ $=$ _____

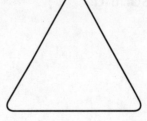

2. $36 \div 9 =$ _____

   $9 \times$ _____ $=$ _____

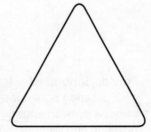

What is the unknown-factor problem for the division equation?

3. $42 \div 6 = ?$

_____ $\times ? =$ _____

4. $? = 18 \div 6$

_____ $=$ _____ $\times ?$

5. $? = 72 \div 9$

_____

6. $14 \div 2 = ?$

_____

7. Kiora arranges a museum collection of 28 bugs in 4 equal rows in a display case. How many bugs are in each row? Explain.

8. Hye studies a new language 6 days a week, for a total of 54 minutes. He studies the same amount of time each day. How many minutes does he study each day? Explain.

9. What 4 equations can you write using the fact triangle? Justify your reasoning.

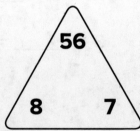

Provide opportunities for your child to practice using unknown factors in multiplication equations to find the quotient in a division equation. Give him or her a division equation. Have him or her write a related multiplication problem and identify the unknown factor to find the quotient.

Additional Practice

Name _____

Review

You can use the relationship between multiplication and division to help you divide by 5 and 10.

A nickel has a value of 5 cents. Use multiplication and division to find how many nickels are in 45 cents.	A dime has a value of 10 cents. Use multiplication and division to find how many dimes are in 30 cents.

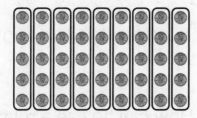

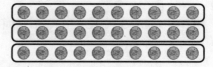

$$45 \div 5 = \mathbf{9}$$

$$5 \times \mathbf{9} = 45$$

So, 9 nickels equal 45 cents.

$$30 \div 10 = \mathbf{3}$$

$$10 \times \mathbf{3} = 30$$

So, 3 dimes equal 30 cents.

---

1. Which multiplication equation can help you find the quotient for $35 \div 5 = ?$ Choose all that apply.

   A. $? \times 5 = 35$ **B.** $35 = 5 \times ?$ **C.** $? = 35 \times 5$ **D.** $5 \times ? = 35$

2. Which multiplication equation can help you find the quotient for $70 \div 10 = ?$ Choose all that apply.

   A. $10 = 70 \times ?$ **B.** $70 = 10 \times ?$ **C.** $? \times 10 = 70$ **D.** $10 \times ? = 70$

How can you complete the equation to make it true?

3. $20 \div$ _____ $= 5$

4. _____ $= 40 \div 5$

5. _____ $\div 3 = 10$

6. $60 \div 10 =$ _____

7. Complete each equation with a number in a box that makes the equation true.

$80 \div 10 =$ ☐   5

$25 \div 5 =$ ☐   2

$10 \div 10 =$ ☐   8

$30 \div 5 =$ ☐   1

$20 \div 10 =$ ☐   3

$15 \div 5 =$ ☐   6

8. How does knowing the fact $5 \times 4 = 20$ help you find $20 \div 5$?

9. Shang is studying for a math test. How can he use a related multiplication fact to help him find the unknown divisor in $60 \div ? = 10$?

10. Natalie has 35 stickers. She places stickers in equal groups on 7 sheets of paper. How can you use an equation to show the number of pages Natalie uses?

Math @ Home Activity

Look for packages or contains that hold objects that are multiples of 5 or 10. Have your child separate the objects into 5 or 10 groups. Determine how many objects are in each group.

Additional Practice

Name _____

Review

You can use related multiplication and division facts to divide by 0 and 1.

When the dividend is 0, the quotient is always 0. Zero multiplied by any number is 0. $0 \div 5 = \mathbf{0}$ $5 \times \mathbf{0} = 0$	A group of objects cannot be divided into 0 groups. Any number multiplied by 0 will always be 0. $5 \div 0$ - It can't be done. $0 \times ? = 0$
When you divide any number by 1, the quotient is the same as the dividend. Any number multiplied by 1 equals itself. $5 \div 1 = \mathbf{5}$ $1 \times \mathbf{5} = 5$	When you divide a number by itself, the quotient is always 1. One multiplied by any number equals the number. $5 \div 5 = \mathbf{1}$ $5 \times \mathbf{1} = 5$

How can you complete the equation to make it true? Cross out any equation that cannot be solved.

1. _____ $= 0 \div 10$

2. _____ $\div 1 = 2$

3. $7 \div 7 =$ _____

4. $8 \div 0 =$ _____

5. $8 =$ _____ $\div 1$

6. $0 =$ _____ $\div 9$

7. $7 \div 0 =$ _____

8. $3 \div$ _____ $= 1$

9. $4 =$ _____ $\div 1$

10. $0 \div 4 =$ _____

11. Choose *True* or *False* for each statement.

	True	False
When you divide a number by itself you get 1.		
When you divide a number by 0 the quotient is 0.		
When you divide a number by 1 the quotient is 1.		
When the dividend is 0 the quotient is 0.		

12. Zack and Stacey want to share a container of strawberries. They open the container and realize that there are no strawberries left. How many strawberries do Zack and Stacey get? Explain.

13. There are 6 dogs and 6 dog crates. The dogs are divided evenly among the crates. How many dogs are in each crate? Write a related multiplication and division equation to justify your answer.

14. There are 10 fish at a pet store. There is 1 fish in each fish bowl. How many fish bowls are there? Write a related multiplication and division equation to justify your answer.

Math @ Home Activity

Provide you child division with 1 and 0. Give him or her real-world problems where 1 is the divisor or the quotient, and where 0 is the dividend and the quotient.

Additional Practice

Name _____

Review

You can use related multiplication facts to divide by 3 and 6.

| You can use an array to determine the unknown. | You can use a fact triangle to determine the unknown. |

You can use an array to determine the unknown.

$12 \div 3 = ?$

$3 \times ? = 12$

3 rows of 4, 12 counters in all

$3 \times 4 = 12$

$12 \div 3 = 4$

You can use a fact triangle to determine the unknown.

$12 \div 6 = ?$

$? \times 6 = 12$

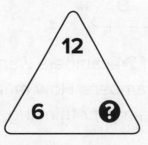

$2 \times 6 = 12$

$12 \div 6 = 2$

How can you use the known facts to complete the equations?

$3 \times 6 = 18$	$3 \times 9 = 27$	$3 \times 10 = 30$
$6 \times 6 = 36$	$6 \times 7 = 42$	$6 \times 8 = 48$

1. $18 \div 3 =$ _____

2. _____ $= 42 \div 6$

3. _____ $= 30 \div 3$

4. $48 \div 6 =$ _____

5. $27 \div 3 =$ _____

6. _____ $= 36 \div 6$

Find the unknown number in the fact triangle. Then, write the four related facts.

7.

_____   _____

_____   _____

8.

_____   _____

_____   _____

9. A group of 24 campers wants to rent cabins. Each cabin sleeps 3 campers. How many cabins should the campers rent? Explain your thinking.

10. A band director arranges chairs in the band room. She has 60 chairs to arrange in 6 rows. How many chairs will be in each row? Explain your thinking.

Math @ Home Activity

Look for situations around your home where you can ask your child to show how to divide by 3 and 6 using related multiplication facts.

Additional Practice

Name _____

Review

You can use the multiplication fact table to show that related multiplication facts can be used to divide by 9.

Find the product in the row or column labeled with the known factor. The unknown factor is the label for the row or column where the product is found.

$45 \div 9 = ?$

$9 \times ? = 45$

$9 \times 5 = 45$

$45 \div 9 = 5$

X	0	1	2	3	4	5	6	7	8	9	10
0	0	0	0	0	0	0	0	0	0	0	0
1	0	1	2	3	4	5	6	7	8	9	10
2	0	2	4	6	8	10	12	14	16	18	20
3	0	3	6	9	12	15	18	21	24	27	
4	0	4	8	12	16	20	24	28	32		40
5	0	5	10	15	20	25	30	35		45	50
6	0	6	12	18	24	30	36	42	48	54	60
7	0	7	14	21	28	35	42	49	56	63	70
8	0	8	16	24	32	40	48	56	64	72	80
9	0	9	18	27	36	45	54	63	72	81	90
10	0	10	20	30	40	50	60	70	80	90	100

How can you use the multiplication fact table to write a related multiplication fact and find the unknown quotient?

1. $90 \div 9 =$ _____

$9 \times$ _____ $=$ _____

2. $72 \div 8 =$ _____

$8 \times$ _____ $=$ _____

3. $18 \div 9 =$ _____

$9 \times$ _____ $=$ _____

4. $63 \div 9 =$ _____

$9 \times$ _____ $=$ _____

5. $36 \div 9$ _____

$9 \times$ _____ $=$ _____

6. $81 \div 9 =$ _____

$9 \times$ _____ $=$ _____

How can you complete the equation to make it true?

7. _____ = 54 ÷ 9

8. 45 ÷ 9 = _____

9. _____ = 9 ÷ 9

10. 27 ÷ 9 = _____

11. What are two related division equations for the multiplication fact?

9 × 1 = 9   _____

_____

12. A total of 72 tourists visit an art museum. The tourists are divided into 9 equal groups. How many tourists are in each group? Justify your answer.

13. Beth finds 27 ÷ 9 = ? using a multiplication fact table. She cannot remember whether she should use the 9s column or the 9s row. Which should she do? Explain.

Math @ Home Activity

Have you child use a multiplication fact table to practice division. Give your child a product of a 9s fact and one factor. Have him or her find the unknown factor on the multiplication fact table.

Additional Practice

Name _____

Review

You can use a multiplication fact table to help divide by 7.

You can use
$28 \div 7 = ?$ and the related
multiplication fact to find
the unknown quotient.

$28 \div 7 = ?$

$7 \times ? = 28$

$7 \times 4 = 28$

$28 \div 7 = 4$

×	0	1	2	3	4	5	6	7	8	9	10
0	0	0	0	0	0	0	0	0	0	0	0
1	0	1	2	3	4	5	6	7	8	9	10
2	0	2	4	6	8	10	12	14	16	18	20
3	0	3	6	9	12	15	18	21	24	27	30
4	0	4	8	12	16	20	24	28	32	36	40
5	0	5	10	15	20	25	30	35	40	45	50
6	0	6	12	18	24	30	36	42	48	54	60
7	0	7	14	21	28	35	42	49	56	63	70
8	0	8	16	24	32	40	48	56	64	72	80
9	0	9	18	27	36	45	54	63	72	81	90
10	0	10	20	30	40	50	60	70	80	90	100

How can you use the multiplication fact table to write a related multiplication fact and find the unknown quotient?

1. $56 \div 7 =$ _____

    $7 \times$ _____

2. $63 \div 7 =$ _____

    $7 \times$ _____ $= 63$

3. $7 \div 7 =$ _____

    $7 \times$ _____ $= 7$

4. $70 \div 7 =$ _____

    $7 \times$ _____ $= 70$

5. $14 \div 7 =$ _____

    $7 \times$ _____ $= 14$

6. $21 \div 7 =$ _____

    $7 \times$ _____ $= 21$

How can you complete the equation to make it true?

7. $35 \div 7 =$ _____   **8.** _____ $= 63 \div 7$

9. _____ $= 42 \div 7$ **10.** $56 \div 8 =$ _____

11. Which equations are true? Choose all that apply.

    A. $7 \div 7 = 0$ **B.** $14 \div 7 = 3$ **C.** $21 \div 2 = 7$

    D. $28 \div 4 = 7$ **E.** $35 \div 7 = 6$ **F.** $42 \div 6 = 7$

12. Taylor colors the same number of pages of a coloring book every day. He colors 49 pages in one week. How many pages does he color each day? Explain.

13. Lola has 28 DVDs. She wants to divide them equally among 7 shelves. How many DVDs should Lola put on each shelf? Justify your answer.

Math @ Home Activity

Have your child use pennies as counters to solve problems involving division by 7. After solving the problem, have him or her write the division equation and a related multiplication equation.

Additional Practice

Name _____

Review

You can multiply by multiples of 10 by using basic facts, place-value understanding, and patterns.

Phil uses 40 beads for each necklace he makes. Phil makes 6 necklaces. How many beads will he use?

Place Value

$6 \times 40 = ?$

6×4 tens $= 24$ tens

So, $6 \times 40 = 240$.

Decompose

$6 \times 40 = ?$

$6 \times 4 \times 10 = ?$

$24 \times 10 = 240$

Phil uses 240 beads.

How can you use place value to multiply?

1. $7 \times 50 = ?$

_____ $\times$ _____ tens

$=$ _____ tens

So, $7 \times 50 =$ _____.

2. $8 \times 30 = ?$

_____ $\times$ _____ tens

$=$ _____ tens

So, $8 \times 30 =$ _____.

3. $7 \times 70 = ?$

_____ $\times$ _____ tens

$=$ _____ tens

So, $7 \times 70 =$ _____.

4. $5 \times 80 = ?$

_____ $\times$ _____ tens

$=$ _____ tens

So, $5 \times 80 =$ _____.

How can you decompose the multiple of 10 to multiply?

5. 4 × 80 = ☐

4 × ___ × 10 = ☐

___ × 10 = ___

6. 9 × 60 = ☐

9 × ___ × 10 = ☐

___ × 10 = ___

---

7. Judy uses 70 buttons for each art project she makes. She makes 8 art projects. How can you decompose the multiple of 10 to find the number of buttons she uses?

8. Ralph uses 40 gallons of water a day to water his garden. How can you use place value to find how many gallons of water he uses for 5 days?

9. What are two multiplication sentences that use a multiple of 10 and have a product of 120?

_____ × _____ = _____

_____ × _____ = _____

Copyright © McGraw-Hill Education

Additional Practice

Name _____

Review

You can find multiplication patterns with factors and products on the multiplication fact table.

The products of 6 × 5 and 5 × 6 are the same.	The product of 6 × 5 is double the product of 3 × 5.

The products of 6 × 5 and 5 × 6 are the same.

X	0	1	2	3	4	5	6
0	0	0	0	0	0	0	0
1	0	1	2	3	4	5	6
2	0	2	4	6	8	10	12
3	0	3	6	9	12	15	18
4	0	4	8	12	16	20	24
5	0	5	10	15	20	25	30
6	0	6	12	18	24	30	36

$6 \times 5 = 30 \qquad 5 \times 6 = 30$

Factors can be multiplied in any order and the product does not change. This is a property of multiplication.

The product of 6 × 5 is double the product of 3 × 5.

X	0	1	2	3	4	5
0	0	0	0	0	0	0
1	0	1	2	3	4	5
2	0	2	4	6	8	10
3	0	3	6	9	12	15
4	0	4	8	12	16	20
5	0	5	10	15	20	25
6	0	6	12	18	24	30

$6 \times 5 = 3 \times 5 + 3 \times 5$

$6 \times 5 = 15 + 15 = 30$

You can decompose 6s facts into two 3s facts using a property of multiplication.

---

1. Which products are Even, and which are Odd?

	Even	Odd
$2 \times 9 = ?$		
$7 \times 3 = ?$		
$8 \times 8 = ?$		
$5 \times 7 = ?$		

2. Why are some products even and some products odd?

Use the multiplication table for 3–5.

3. Setia notices a pattern in the multiplication table and highlights it. How can you explain why the products in the column are the same as the products in the row?

✕	0	1	2	3	4	5	6	7	8	9	10
0	0	0	0	0	0	0	0	0	0	0	0
1	0	1	2	3	4	5	6	7	8	9	10
2	0	2	4	6	8	10	12	14	16	18	20
3	0	3	6	9	12	15	18	21	24	27	30
4	0	4	8	12	16	20	24	28	32	36	40
5	0	5	10	15	20	25	30	35	40	45	50
6	0	6	12	18	24	30	36	42	48	54	60
7	0	7	14	21	28	35	42	**49**	56	63	70
8	0	8	16	24	32	40	48	56	64	72	80
9	0	9	18	27	36	45	54	63	72	81	90
10	0	10	20	30	40	50	60	70	80	90	100

4. Find the products of 2 facts. What pattern do you notice?

5. How do the products of 8s facts relate to products of 4s facts? Explain.

Math @ Home Activity

Have your child create a short story, poem, or song that include the multiplication patterns he or she has learned.

Additional Practice

Name _____

Review

You can represent two-step word problems using bar diagrams and equations with a letter for the unknown.

A flower shop sells bouquets with 8 flowers in each bouquet. Coleman buys 6 bouquets and 14 additional flowers. How many flowers does he buy in all ?

Step 1 Use a bar diagram to represent the total number of flowers in the bouquets. Use a letter to represent the unknown. Write an equation to represent the bar diagram.	**Step 2** Then find the total number of flowers Coleman buys. Use a bar diagram and equation to represent the second step in the equation.
<table><tr><td>8</td><td>8</td><td>8</td><td>8</td><td>8</td><td>8</td></tr></table> ├──────────b──────────┤ *total number of flowers in bouquets* $6 \times 8 = b$ $48 = b$	<table><tr><td>48</td><td>14</td></tr></table> ├──────────f──────────┤ *total number of flowers* $48 + 14 = f$ $62 = f$

What equation is represented by the bar diagram?

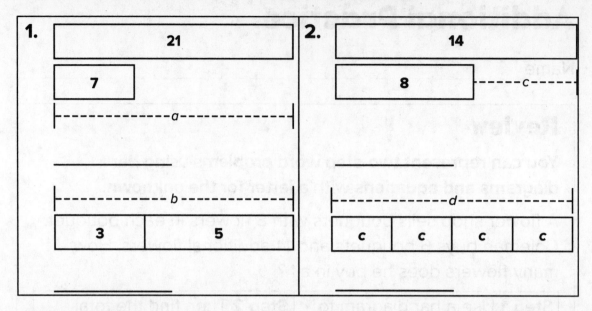

1.
21

7

- - - - - - - - a - - - - - - - -

- - - - - - - - b - - - - - - - -

3	5

2.
14

8	- - - c - - -

- - - - - - - d - - - - - - -

c	c	c	c

How can you use equations with a letter for the unknown to solve the problems?

3. Al needs to make 64 favors for a party. He has already made 10 favors. He has 6 weeks to make the remaining favors. He makes the same number of favors in each of the 6 weeks. How many favors will Al make each week?

4. Mrs. Tice buys pencils in packs of 8. She buys 9 packs and 12 additional pencils. How many pencils does she buy in all?

5. Don divides 45 tickets among 5 friends. He gives each friend 4 more tickets. How many tickets does each friend receive?

6. Steve has 6 boxes of trading cards. There are 6 cards in each box. He buys 11 more cards. How many cards does he have?

Math @ Home Activity

Ask your child to divide a group of paper clips into equal groups and then add or subtract a certain number for each group. Then have him or her write the equations with a letter for the unknown to represent the situation. Repeat the activity with different numbers.

Additional Practice

Name _____

Review

You can use mental math and estimation to determine whether an answer is reasonable.

A pet store has 7 fish tanks with 8 fish in each tank. The store sells 13 fish. Nam thinks the store has 28 fish left. Is his answer reasonable?

You can use mental math.	You can estimate.
$7 \times 8 = f$ $\mathbf{56 = f}$ The store starts with 56 fish.	$56 - 13 = l$ $\downarrow \quad \downarrow$ $55 - 10 = 45$ The pet store has about 45 fish left.
Estimate: 45	Nam's Answer: 28

Nam's answer is not reasonable because it is not close to the estimate.

How can you estimate to determine the reasonableness of an answer? Choose the reasonable answer.

1. Parvati is at school 8 hours a day, 4 days a week. On Wednesdays, she is at school for 6 hours. How many hours does Parvati spend at school each week?

   A. 24 hours   **B.** 38 hours

   C. 48 hours   **D.** 52 hours

How can you determine and explain whether the answer is reasonable?

2. Westly can use the family computer for 295 minutes each week. Five days of the week, she uses the computer 30 minutes each day. She thinks she has 265 minutes of computer time left for Saturday and Sunday.

3. Ms. Gregg buys 7 packages of scissors. There are 6 scissors in each package. Then she buys 11 individual scissors. She thinks she has 53 scissors.

Find the solution. Then show an estimate to check the reasonableness of your answer.

4. Quentin builds 4 robots with his construction blocks set. He needs 80 construction blocks to build one robot. He has 463 construction blocks. He thinks he will have 143 construction blocks left. Is his answer reasonable?

5. Haley has 27 bottles of paint. She buys 2 packages of paint. There are 8 bottles of paint in each package. She thinks she has 35 bottles of paint. Is her estimate close to the answer?

Math @ Home Activity

Help your child make a list of everyday activities that use math. Then have him or her decide when using a reasonable answer would be a good idea and when it would not.

Additional Practice

Name _____

Review

The distance around the outside of a figure is the perimeter. You can find the perimeter of a figure using addition or multiplication.

Opposite sides of a rectangle have equal lengths.	All sides of a square have the same length.

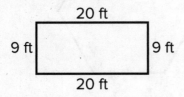

$$20 + 9 + 20 + 9 = 58$$

You can add the side lengths to find the perimeter.

The perimeter is 58 feet.

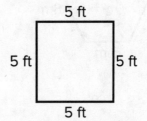

$$5 + 5 + 5 + 5 = 20; 4 \times 5 = 20$$

You can add or multiply the side lengths to find the perimeter.

The perimeter is 20 feet.

How can you complete the equation and find the perimeter of the figure?

1.

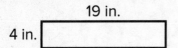

Perimeter = _____ +

_____ + _____ + _____

Perimeter = _____ inches

2.

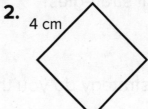

Perimeter = _____ +

_____ + _____ + _____

Perimeter = _____ × _____

Perimeter = _____ centimeters

How can you find the perimeter of the figure? Label with units.

3. _____

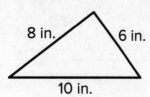
8 in.   6 in.
10 in.

4. _____

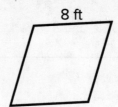
8 ft

5. _____

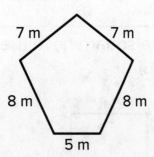
7 m   7 m
8 m   8 m
5 m

6. _____

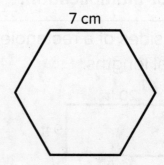
7 cm

7. _____

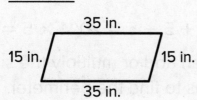
35 in.
15 in.   15 in.
35 in.

8. _____

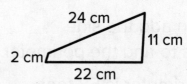
24 cm
11 cm
2 cm
22 cm

9. Ida is building a garden box that has 6 sides.
Each side is 9 feet long.

    a. How can she determine the perimeter using two
different strategies?

    b. Which strategy do you think is more efficient?

Math @ Home Activity

Use a ruler or tape measure to find the dimensions of flat objects around your home. Have your child calculate the perimeter of each object based on the dimensions. Be sure to include the units.

Additional Practice

Name _____

Review

You can use the perimeter to find the unknown side length. You can write addition and subtraction equations to find an unknown side length.

The perimeter of the shape is 51 inches. You can write an addition equation to represent the perimeter.

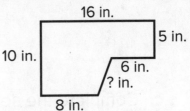

16 in.
5 in.
10 in.
6 in.
? in.
8 in.

$51 = 16 + 5 + 6 + ? + 8 + 10$

Add the side lengths you know and subtract to solve the equation.

$51 = 45 + ?$ ⟶ $51 - 45 = ?$ ⟶ $51 - 45 = 6$

The unknown side length is 6 inches.

Perimeter $= 16 + 5 + 6 + 6 + 8 + 10 = 51$.

---

1. How can you find the unknown side length of the figure? Complete the equations and the sentence.

The perimeter is 49 centimeters.

_____ + _____ + _____ + _____ = _____

Add the known side lengths.

_____ + _____ + _____ = _____

Subtract the known amount.

_____ − _____ = _____

Solve for the unknown.

? = _____

The unknown side length is _____ centimeters.

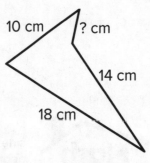

10 cm
? cm
14 cm
18 cm

2. The base of a fish tank is a rectangle with a length of 20 inches. The perimeter is 60 inches.

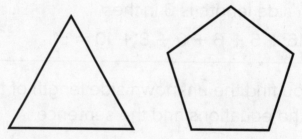

20 in.

? in.   ? in.

20 in.

   a. Write an addition equation to find the unknown width of the base.

   b. Solve. Show your work.

   c. Complete the sentence.

      The unknown width is _____ inches.

3. The triangle and pentagon both have a perimeter of 30 meters. The triangle has sides of equal length, and the pentagon has sides of equal length.

What are the side lengths of each polygon? Explain.

Math @ Home Activity

Have your child use a ruler to draw a polygon and measure all but one of the side lengths. Help him or her label the side lengths to the closest centimeter or other unit. Measure the last side without your child and add all the side lengths to find the perimeter. Give your child the perimeter and have him or her calculate the unknown side length. Then your child can measure the last side of the polygon to check his or her work.

Additional Practice

Name _____

Review

You can solve real-world problems involving perimeter and area. For any rectangle, the perimeter is equal to length + width + length + width, and the area is equal to length × width.

A rectangle has an area of 16 square units. What could be the length and width of the rectangle?

Since the area of a rectangle is equal to length × width, find a factor pair for 16. There are three possibilities: 1 and 16, 2 and 8, and 4 and 4. So, the dimensions of the rectangle could be 1 unit by 16 units, 2 units by 8 units, or 4 units by 4 units.

What is the area and perimeter of each rectangle? Include the unit used for measuring.

1.

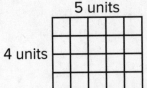
5 units
4 units

Area = _____

Perimeter = _____

2.

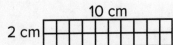
10 cm
2 cm

Area = _____

Perimeter = _____

3.

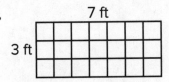
7 ft
3 ft

Area = _____

Perimeter = _____

4.

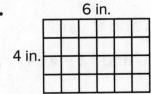
6 in.
4 in.

Area = _____

Perimeter = _____

5. Rectangles A, B, and C have the same area: 18 square inches. What could be their lengths and widths? Include units.

Rectangle A   Rectangle B

Rectangle C

Rectangle A: Length = _____

Width = _____

Rectangle B: Length = _____

Width = _____

Rectangle C: Length = _____

Width = _____

6. A rectangular cement pad has an area of 30 square feet. What could be the possible side lengths and perimeter of the cement pad? Explain your thinking.

7. A rectangular patio has a perimeter of 24 yards. What could be the side lengths of the patio? Choose all that apply.

A. 1 yard and 11 yards

B. 3 yards and 9 yards

C. 4 yards and 8 yards

D. 12 yards and 12 yards

Math @ Home Activity

Write the numbers 1 through 30 on small pieces of paper, and place them in a hat. Have your child select one piece. Draw two or three rectangles with the area of the number selected. Then determine the perimeter of each rectangle. Repeat with other numbers.

Additional Practice

Name _____

Review

You can solve problems involving length measurements using different strategies.

Curt needs 9 inches of string to tie each tomato plant up. He has 5 tomato plants. How many inches of string does he need?

You can represent the problem with a multiplication equation.

$$5 \times 9 = c$$

You can use a bar diagram.	You can decompose a factor.
|----------- c -----------| | 9 | 9 | 9 | 9 | 9 | $5 \times 9 = c$ $45 = c$	$5 \times \mathbf{9} = 5 \times \mathbf{5} + 5 \times \mathbf{4}$ $5 \times 9 = 25 + 20$ $5 \times 9 = 45$

Curt needs 45 inches of string.

What number completes the equation?

1. $6 \times 6 =$ _____

2. _____ $= 40 \div 10$

3. _____ $\div 8 = 0$

4. $8 \times$ _____ $= 48$

5. $63 \div 9 =$ _____

6. _____ $\times 5 = 15$

7. $2 \times$ _____ $= 20$

8. _____ $\div 1 = 1$

9. Evelyn sews together a row of 10 quilt squares. The total length of the row is 50 inches. How can you represent the problem and find the length of each square?

10. Lee is making holes in her garden to plant different vegetables. For each vegetable, the number of plants, the length of planting space for each plant, and the total number of inches of garden space she needs changes. What is the unknown for each vegetable?

Number of Plants	Length of Planting Space (inches)	Total Amount of Planting Space (inches)
6	8	_____
2	_____	20
_____	3	27
4	9	_____

11. Geoffrey has wood planks that are 4 feet long. He lays 7 planks end to end. How can you represent the problem and find the total length of the wood planks?

12. Parker needs to find the product of 6 × 7. She creates an array with 6 rows and 7 columns. She says she can double the total number of objects in the array to find the product of 6 × 7. How do you respond to Parker?

Math @ Home Activity

Provide opportunities for your child solve problems involving measurement. Place cards that are the same length end-to-end. Tell your child the total length of the cards or the length of one card. Then have him or her find the length of one card or the total length of all the cards.

Additional Practice

Name _____

This material may be reproduced for licensed classroom use only and may not be further reproduced or distributed.

Review

You can use measuring cups to measure amounts of liquid.
Liquids can be measured in milliliters (mL) or liters (L).

1 liter = 1,000 milliliters

Use liters to measure large amounts of liquids.

Use milliliters to measure small amounts of liquid.

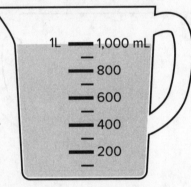

1L — 1,000 mL
— 800
— 600
— 400
— 200

1 liter of liquid

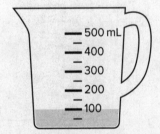

— 500 mL
— 400
— 300
— 200
— 100

100 milliliters of liquid

The amount of liquid in a container is called liquid volume.

What is the liquid volume in the container?

1.

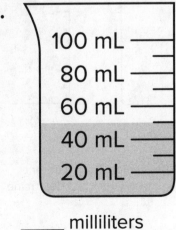

100 mL —
80 mL —
60 mL —
40 mL —
20 mL —

_____ milliliters

2.

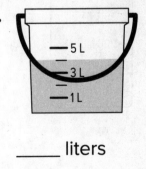

— 5 L
— 3 L
— 1 L

_____ liters

3.

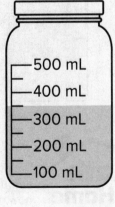

— 500 mL
— 400 mL
— 300 mL
— 200 mL
— 100 mL

_____ milliliters

4. Nan fills her water bottle with 950 milliliters of water. How can you shade the water bottle to represent the liquid volume of the water?

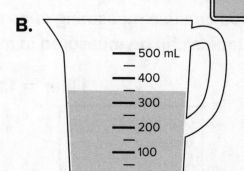

Which measuring cup shows the given liquid volume?

5. 400 milliliters

A.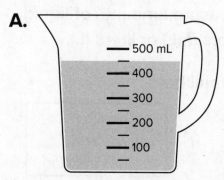

B.

C.

D.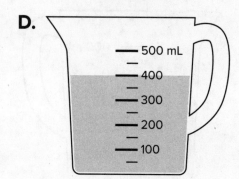

6. 3 liters

A. 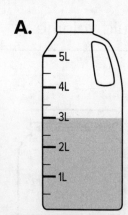 **B.** 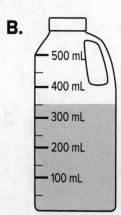 **C.** 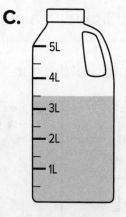 **D.**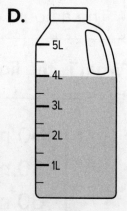

Math @Home Activity

Provide opportunities for your child to measure liquids. Pour different amounts of water into a measuring cup. Then, have your child determine how much water is in the measuring cup.

Additional Practice

Name _____

Review

Mass is the amount of matter in an object.

You measure mass in **grams (g)** and **kilograms (kg)**.

1,000 grams (g) is equal to 1 kilogram (kg).

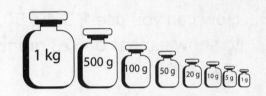

Find the mass of the apple. Place the apple on one side of the scale and weights on the other until they are level.

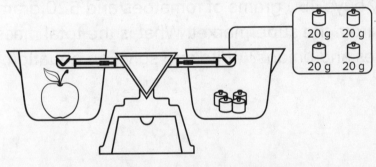

$$20 + 20 + 20 + 20 = 80$$

The mass of the apple is 80 grams.

---

1. What is the mass of the baseball? _____ grams

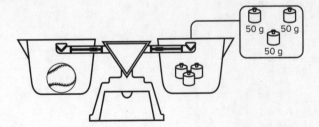

2. What is the mass of the lemon? _____ grams

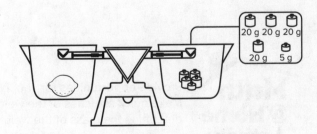

3. The weights show the mass of a cat. What is the mass of the cat?

_____ kilograms.

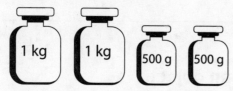

4. The weights show the mass of a golf ball. What is the mass of the golf ball?

_____ grams

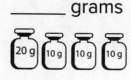

5. How can you use 1, 5, 10, 50, 100, or 500-gram weights to show a mass of 623 grams? Write an equation.

6. Brittany buys 450 grams of tomatoes and 520 grams of onions from the supermarket. What is the total mass of the tomatoes and onions? Write and solve an equation.

7. There are 7 orange slices in a bowl. Each slice has the same mass. The orange slices have a total mass of 35 grams. What is the mass of each orange slice?

Provide opportunities for your child to work with the mass of an object. Make quick sketches of the weights on Page 1. Then have your child determine the mass of the weight you sketch.

Additional Practice

Name

Review

You can use everyday objects to help you estimate the mass of an object. Then, you can solve problems about mass using equations and strategies.

Use everyday objects to estimate	Paper Clip	Apple	Dictionary
	1 gram (g)	100 grams (g)	1 kilogram (kg)

Which is the best estimate for the mass of the object?

1. Camel

A. 550 grams

B. 550 kilograms

C. 50 grams

D. 50 kilograms

2. Cat

A. 30 grams

B. 30 kilograms

C. 3 grams

D. 3 kilograms

3. Golf ball

A. 5 grams

B. 5 kilograms

C. 50 grams

D. 50 kilograms

4. A cookie recipe uses 90 grams of sugar and 210 grams of flour. What is the difference in the mass of the sugar and flour? Write an equation and solve for the unknown.

5. Ernest feeds his dog 3 times a day. Each time he gives his dog 500 grams of dog food. How much dog food does he use in one day? Write an equation and solve for the unknown?

6. Zoey has some powdered drink mix with a mass of 36 grams. She distributes the same amount of powdered drink mix to 9 glasses. What is the mass of the powdered drink mix in each glass? Write an equation and solve for the unknown.

7. Yvonne says a lunch box would have about the same mass as 7 bananas. Do you agree? Explain your answer.

1 gram

100 grams

Additional Practice

Name _____

Review

You can tell time to the nearest minute by looking at the hour and minute hands on a clock.

The clock shows the time Max's bus is scheduled to arrive. How can he know the exact time?

When writing or telling time to the nearest minute, you start with the hour hand.

The hour hand is between 4 and 5, indicating that the hour is 4.

Next, look at the minute hand. The minute hand is past the 5, meaning it is at least 25 minutes past 9. It is 2 tick marks past the 5, indicating Max's bus arrives at 27 minutes past 4, or 4:27.

You can also say that Max's bus arrives at 33 minutes to 5.

How can you write the time shown on each clock?

1.

____:____

2.

____:____

3.

____:____

4. Alex, Aries, and Axel got on an airplane at different times.

a. Write the time each person got on the airplane.

Alex

Aries

Axel

_____:_____   _____:_____   _____:_____

b. What would Aries's clock look like if she got on the airplane at 8:52?

Aries

---

5. Rachael began mowing the lawn at 1:36.

a. Circle the clock that shows the time Rachael began mowing the lawn.

b. What time does the clock that you did *not* circle show?

_____

6. Carmen and Zeke left the library at different times. Who left last? Explain your answer.

Carmen

Zeke

Math @ Home Activity

Practice telling time to the nearest minute with your child. For example, when doing different activities throughout the evening, such as eating dinner or brushing teeth, ask your child to tell you the time to the nearest minute. Encourage him or her to tell you the time in two different ways.

Additional Practice

Name _____

Review

You can use data from a table to create a scaled picture graph that uses the same picture to represent more than one type of data. A key shows the value of each picture.

Favorite Drink	
Drink	**Number of Students**
Water	20
Milk	5
Apple Juice	10
Orange Juice	15

You can use the key and the number of pictures in each category to calculate the number of students in each category.

Water: 4 × 5 students = 20 students

Milk: 1 × 5 students = 5 students

Apple Juice: 2 × 5 students = 10 students

Orange Juice: 3 × 5 students = 15 students

1. Use the picture graph to answer questions about the number of students from each grade that are in the school band.

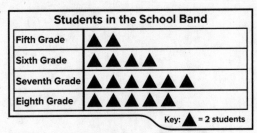

   a. How many eighth-grade students are in the band? _____ students

   b. Which grade has the most number of students in the band? _____ grade

Use the picture graph to answer questions about the number of students from each grade that are in the school band.

   c. How many students are in the grade that has the most number of students in the band? _____ students

2. The table shows the results of a survey about students' favorite sports. Complete the scaled picture graph.

Favorite Sport	
Sport	Number of Students
Baseball	12
Football	24
Volleyball	6
Soccer	15

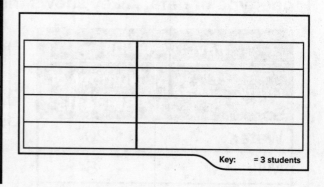

Key: = 3 students

   a. How many pictures did you use in the volleyball category? Explain your answer.

   b. What is a different way you could scale the picture graph to represent the data shown in the table Favorite Sport?

Math @ Home Activity

Look for situations around your home where your child can practice creating scaled picture graphs. For example, have your child ask everyone in the family their favorite fruit. Your child can then create a graph using that data. Family members can ask and answer questions relating to the data in the graph.

Additional Practice

Name _____

Review

You can create a scaled bar graph that uses a scale that increases by a number greater than one.

Favorite Season	
Season	**Number of Kids**
Spring	8
Summer	6
Fall	10
Winter	2

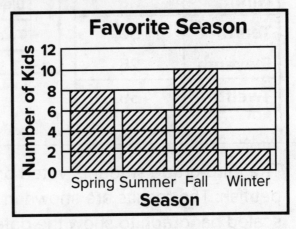

Think about the numbers being represented. Create a scale, or the amount each interval represents.

On this graph, the scale is 2. So, each interval represent 2 kids.

You can use vertical or horizontal bar graphs to represent large amounts of data.

How can you complete the bar graph using the data in the table?

1.

Favorite Pet	
Pet	**Number of Family Members**
Cat	2
Dog	10
Fish	8
Hamster	6

2. The table shows the number of students who walk to school at Brooke High School. Complete the horizontal bar graph using the data in the table.

Walkers	
Grade	**Number of Students**
Eighth	30
Ninth	20
Tenth	50
Eleventh	25
Twelfth	35

Number of Walkers

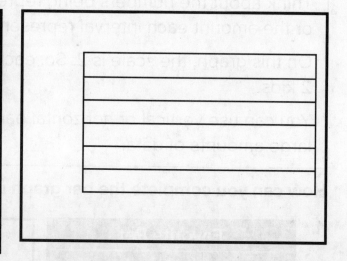

3. The third-grade students voted on a name for their classroom pet fish. The results are shown in the data table. Draw a scaled bar graph to show the data. Be sure to add a title and complete the scale.

Name for Classroom Fish	
Name	**Number of Votes**
Fred	4
Frank	10
Fae	12
Faye	14

Math @ Home Activity

Have your child record the number of different animals or plants he or she sees while out on a walk, riding in the car, etc. Your child can then create a scaled bar graph using the data. Have him or her write an equation that shows the total number of animals/plants seen.

Additional Practice

Name _____

Review

You can use an inch ruler to measure lengths. You can measure to a more precise measurement than the inch marks by using the half-inch and the quarter-inch tick marks on a ruler.

The half-inch tick marks partition each inch into two equal parts. The quarter-inch tick marks partition each inch into four equal parts.

Measure the bee to the nearest quarter inch.

The bee is between 1 and 2 whole inches. The front of the bee is at the $\frac{1}{4}$ tick mark. So, the bee is $1\frac{1}{4}$ inches long.

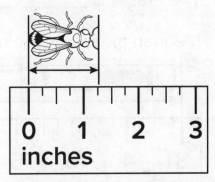

How can you use an inch ruler to measure each object to the nearest half inch?

1.

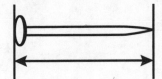

The nail is

_____ inches long.

2.

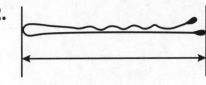

The bobby pin is

_____ inches long.

How can you use an inch ruler to measure each object to the nearest quarter inch?

3.

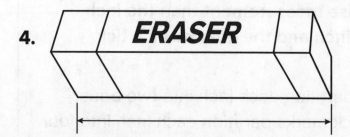

The bracelet is _____ inches long.

4.

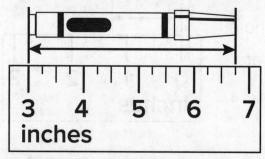

The eraser is _____ inches long.

5. Meghan measures the marker as shown.

a. What could Meghan do to make measuring the marker easier?

b. What is the length of the marker? _____ inches

Math
@ Home
Activity

Give your child many opportunities to practice measuring objects to the nearest inch and the nearest quarter inch. For example, while making a craft, have your child measure each part of the craft to the nearest half inch or quarter inch. Be sure he or she lines up each object with the 0-inch mark.

Additional Practice

Name _____

Review

You can create a line plot using a given data set.

Begin by placing the measurements on a tally chart.

Draw and partition a number line according to the data values. Since the data values are to the nearest quarter inch, partition each whole into fourths.

Insect Lengths (in.)			
2	4	2	$2\frac{3}{4}$
$3\frac{1}{2}$	4	$3\frac{1}{4}$	2
3	$3\frac{1}{4}$	3	$2\frac{1}{4}$

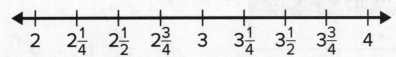

Then, place Xs above each tick mark for each data value. Be sure to include a title for the line plot.

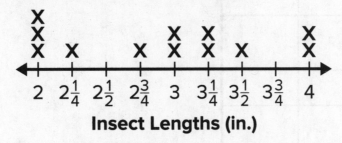

Insect Lengths (in.)

Zack measures the lengths of the rocks in his rock collection.

1. How many rocks are in Zack's collection?

_____ rocks.

Lengths of Rocks (in.)

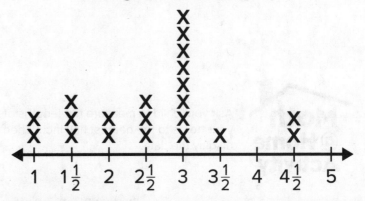

2. Brock measures his pencils to the nearest half inch. He records the measurements in a table.

Pencil Length (in.)			
1	2	$2\frac{1}{2}$	3
3	$1\frac{1}{2}$	$3\frac{1}{2}$	2
2	3	$3\frac{1}{2}$	1

a. How can you create a tally chart from the data on the table?

b. How can you create a line plot from the data in the table?

Lengths of Pencils	
Length (in.)	Tally

Lengths of Pencils (in.)

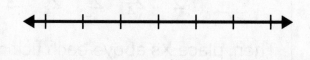

Math @ Home Activity

Ask your child to measure the lengths of the items in a collection around your home to the nearest half inch. Then have him or her create a line plot to display the data.

Lesson **13-1**

Additional Practice

Name _____

Review

A polygon is a closed 2-dimensional shape formed by three or more straight sides that do not cross. You can name a polygon based on how many sides it has.

Triangle	Quadrilateral	Pentagon	Hexagon	Octagon
3 sides 3 angles	4 sides 4 angles	5 sides 5 angles	6 sides 6 angles	8 sides 8 angles

---

Is the shape a polygon? Explain why or why not.

1.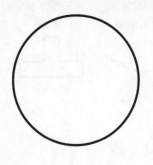

2.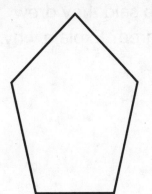

3.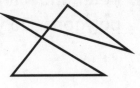

How can you classify the polygon?

4. _____

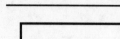

5. _____

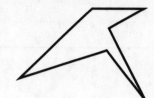

6. _____

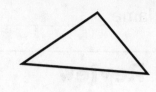

7. _____

8. _____

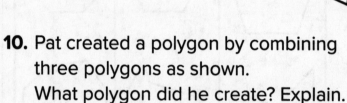

9. _____

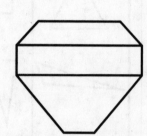

10. Pat created a polygon by combining three polygons as shown.
What polygon did he create? Explain.

11. Peter and Maria both said they drew polygons. Do you agree? Explain why.

Maria's Figure   Peter's Figure

Additional Practice

Name _____

Review

You can describe a quadrilateral by thinking about the number of right angles and pairs of equal opposite side lengths it has.

Side Lengths	Right Angles

Side Lengths

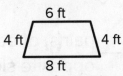

6 ft
4 ft   4 ft
8 ft

1 pair of opposite sides are equal.

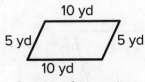

10 yd
5 yd   5 yd
10 yd

2 pairs of opposite sides are equal.

Right Angles

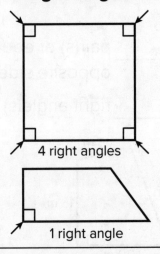

4 right angles

1 right angle

---

1. I am a quadrilateral with, 0 pairs of equal opposite sides, and 2 right angles. Which shape am I?

A.

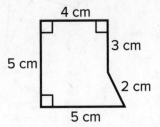

4 cm
5 cm
3 cm
2 cm
5 cm

B.

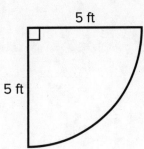

5 ft
5 ft

C.

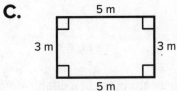

5 m
3 m   3 m
5 m

D.

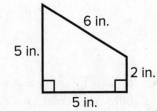

6 in.
5 in.
2 in.
5 in.

How can you describe the quadrilateral using the number of right angles and pairs of equal opposite sides?

2. _____ pair(s) of equal opposite sides

_____ right angle(s)

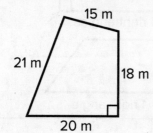

3. _____ pair(s) of equal opposite sides

_____ right angle(s)

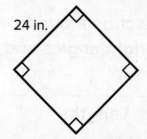

4. _____ pair(s) of equal opposite sides

_____ right angle(s)

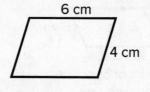

5. _____ pair(s) of equal opposite sides

_____ right angle(s)

6. How can you describe the shape of the kite using its attributes?

Provide opportunities for your child to name attributes of a quadrilateral. Show your child a 4-sided figure. Then have him or her tell you the number of equal sides, and right angles it has.

Additional Practice

Name _____

Review

You can classify quadrilaterals based on their attributes.

Shape	Attributes
rectangle	a quadrilateral with four right angles
rhombus	a quadrilateral with four sides of equal length
square	a quadrilateral with four right angles and four sides of equal length

This quadrilateral is a rectangle because it has four right angles. It is a rhombus because all four sides are equal lengths. Most specifically, it is a square because it is a quadrilateral with four right angles and four sides of equal length.

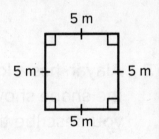

---

1. How can you classify the quadrilaterals shown? Explain.

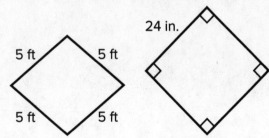

2. How can you classify the quadrilaterals shown? Explain.

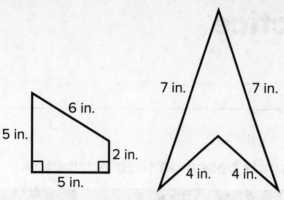

3. Alayah has a keychain with the shape shown. How can you describe the shape of the keychain? Explain.

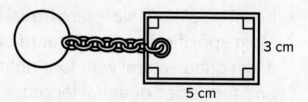

Math @Home Activity

Embark on a quadrilateral safari with your child. Search for at least one example of each type of quadrilateral in your home or community. Discuss which quadrilaterals are more difficult to find.

Additional Practice

Name _____

Review

Your can draw quadrilaterals based on their attributes. The attributes of a quadrilateral include the number of right angles and pairs of equal opposite side lengths.

How can you draw a quadrilateral with the following attributes?

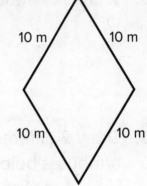

10 m   10 m

10 m   10 m

- no right angles
- all sides are the same length

It cannot be a rectangle because it has no right angles.

It is a rhombus because all sides are the same length.

Based on the attributes, the quadrilateral that you should draw is a rhombus.

How can you draw a quadrilateral to match the description? Some may have more than one answer.

1. 4 right angles, and 2 pairs of equal opposite sides

2. No right angles, 4 sides of equal length

How can you draw a quadrilateral to match the description?

3. 2 pairs of equal opposite sides and 4 right angles

4. 4 sides of equal length, 4 right angles

5. 2 pairs of equal opposite sides and 0 right angles

6. Draw 2 different types of quadrilaterals that match the attributes below.

- no right angles
- two or more sides of the same length

Math @ Home Activity

Look at different pictures of quadrilaterals. Describe the attributes to your child without him or her seeing the picture. Have your child draw the quadrilateral based on the description of the attributes.